高职高专"十三五"规划教材

机电专业英语

宋飞燕　隋瑞歆　李晓燕　主　编
吕世芳　王丽萍　钟学奎　副主编

化学工业出版社

·北京·

本书主要内容包括机电一体化技术、电气技术、传感检测技术、气动和电气动技术、自动控制技术 5 个单元，共 17 节课程，并对科技英语的翻译进行了讲解，每课后有本课文章的翻译参考。所选内容难度适宜、精炼、实用。本书注重扩充自动化技术、机电一体化技术专业英语词汇量，提高学生和技术人员阅读和翻译科技英语资料的能力，扩展和深化对本学科关键技术的认识，增加对专业英语的实际应用能力。为方便教学，配套电子课件。

本书可作为高职高专院校机电一体化专业、自动化专业教学用书，同时也适合中职及应用型本科院校相关专业使用，还可供相关技术人员作为培训用书或学习参考资料。

图书在版编目（CIP）数据

机电专业英语/宋飞燕，隋瑞歆，李晓燕主编. —北京：化学工业出版社，2017.5（2021.5重印）
高职高专"十三五"规划教材
ISBN 978-7-122-29337-4

Ⅰ.①机… Ⅱ.①宋…②隋…③李… Ⅲ.①机电工程-英语-高等职业教育-教材 Ⅳ.①TH

中国版本图书馆 CIP 数据核字（2017）第 060979 号

责任编辑：韩庆利
责任校对：边　涛　　　　　　　　　　　　　　装帧设计：史利平

出版发行：化学工业出版社（北京市东城区青年湖南街 13 号　邮政编码 100011）
印　　装：北京盛通商印快线网络科技有限公司
787mm×1092mm　1/16　印张 9¼　字数 197 千字　2021 年 5 月北京第 1 版第 2 次印刷

购书咨询：010-64518888　　　　　　　　售后服务：010-64518899
网　　址：http://www.cip.com.cn
凡购买本书，如有缺损质量问题，本社销售中心负责调换。

定　　价：26.00 元　　　　　　　　　　　　　　　　版权所有　违者必究

  自动化技术是一门综合性技术，它和控制论、信息论、系统工程、计算机技术、电子学、液压气压技术、自动控制等都有着十分密切的关系，广泛用于工业、农业、国防、科学研究、交通运输、商业、医疗以及家庭等各方面。自动化在人类社会中的地位至关重要，是一个国家或社会现代化水平的重要标志。我国自动化、机电一体化技术科技水平发展日新月异，但与发达国家还存在差距，我们需要学会使用英语这门目前国际上通用的语言，来与其他国家的技术人员进行交流，还需要读懂各种英文资料、文献，以更好地将自动化技术、机电一体化技术应用到生产生活中。故以自动化技术、机电一体化技术为背景的专业英语已成为科研和学习的重要工具。

  编写本书的主要目的是扩充自动化技术、机电一体化技术专业英语词汇量，提高学生和技术人员阅读和翻译科技英语资料的能力，扩展和深化对本学科关键技术的认识，增加对专业英语的实际应用能力。本书主要内容包括机电一体化技术、电气技术、传感检测技术、气动和电气动技术、自动控制技术 5 个单元，共 17 节课程。内容组织系统、精炼、实用，符合高职教育应用性特点。

  本书由包头轻工职业技术学院宋飞燕、隋瑞歆、李晓燕、吕世芳、王丽萍、钟学奎、程秀玲、张燕、刘彦超、付志勇编写。本书英语素材的提供得到了 FESTO（中国）北京分公司高鹏和上海分公司高海华的大力支持，在此向他们深表谢意！本书图片的处理，得到了张宏伟的支持，这里一并感谢！

  本书配套电子课件，可赠送给用本书作为授课教材的院校和老师，可发送邮件至 hqlbook@126.com 索取。

  由于学识有限，书中误漏之处难免，望广大读者批评指正。

编者

Contents

Unit 1 Electromechanical Introduction 1

Lesson 1   What Is Mechatronics? 1
    Part 1   Text 1
    Part 2   New Words and Phrases 2
    Part 3   Technical Words and Phrases 3
    Part 4   Translations 3
    Part 5   Reference Version 4
Lesson 2   Introduction of Modern Manufacturing Technology 6
    Part 1   Text 6
    Part 2   New Words and Phrases 7
    Part 3   Technical Words and Phrases 7
    Part 4   Translations 8
    Part 5   Reference Version 9
Section Ⅰ of Translating Skills：科技英语文体的特点与科技英语翻译概述 11

Unit 2 Electrical Technology 13

Lesson 3   Direct Current and Alternating Current 13
    Part 1   Text 13
    Part 2   New Words and Phrases 15
    Part 3   Technical Words and Phrases 15
    Part 4   Translations 16
    Part 5   Reference Version 17
Lesson 4   Resistor, Capacitor and Inductor 19
    Part 1   Text 19
    Part 2   New Words and Phrases 21
    Part 3   Technical Words and Phrases 21
    Part 4   Translations 22
    Part 5   Reference Version 23
Section Ⅱ of Translating Skills：科技英语中词的翻译 25
Lesson 5   Ohm's Law and Measurement in Electrical Circuits 29

Part 1 Text	29
Part 2 New Words and Phrases	33
Part 3 Technical Words and Phrases	34
Part 4 Translations	35
Part 5 Reference Version	36

Unit 3 Sensor Detection Techniques — 41

Lesson 6 Introduction of Sensor Technology	41
Part 1 Text	41
Part 2 New Words and Phrases	43
Part 3 Technical Words and Phrases	44
Part 4 Translations	44
Part 5 Reference Version	45
Section Ⅲ of Translating Skills：科技英语中长句的翻译	47
Lesson 7 Capacitive and Inductive Proximity Sensors	51
Part 1 Text	51
Part 2 New Words and Phrases	52
Part 3 Technical Words and Phrases	52
Part 4 Translations	53
Part 5 Reference Version	54
Lesson 8 Ultrasonic Sensor	55
Part 1 Text	55
Part 2 New Words and Phrases	58
Part 3 Translations	58
Part 4 Reference Version	59
Section Ⅳ of Translating Skills：被动语态在科技英语中的应用与翻译	62
Lesson 9 Temperature Sensor	65
Part 1 Text	65
Part 2 New Words and Phrases	67
Part 3 Technical Words and Phrases	67
Part 4 Translations	68
Part 5 Reference Version	69
Section Ⅴ of Translating Skills：定语从句在科技英语中的应用与翻译	71

Unit 4 Pneumatics and Electropneumatics — 75

Lesson 10 Introduction of Pneumatics and Electropneumatics	75
Part 1 Text	75

Part 2	New Words and Phrases	76
Part 3	Technical Words and Phrases	78
Part 4	Translations	78
Part 5	Reference Version	79

Lesson 11 Basic Control Engineering Terms 81
    Part 1　Text 81
    Part 2　New Words and Phrases 83
    Part 3　Technical Words and Phrases 85
    Part 4　Translations 85
    Part 5　Reference Version 86

Lesson 12 Pneumatic and Electropneumatic Control Systems 89
    Part 1　Text 89
    Part 2　New Words and Phrases 91
    Part 3　Technical Words and Phrases 91
    Part 4　Translation 92
    Part 5　Reference Version 93

Section Ⅵ of Translating Skills：家用电器、电子产品说明书的翻译 96

Unit 5 Automatic Control Techniques　　103

Lesson 13 The Introduction to Controls 103
    Part 1　Text 103
    Part 2　New Words and Phrases 106
    Part 3　Technical Words and Phrases 108
    Part 4　Translations 109
    Part 5　Reference Version 110

Lesson 14 The Programmable Logical Controller Techique 113
    Part 1　Text 113
    Part 2　New Words and Phrases 115
    Part 3　Technical Words and Phrases 115
    Part 4　Translations 116
    Part 5　Reference Version 117

Lesson 15 Open Loop Control System and Closed Loop Control 119
    Part 1　Text 119
    Part 2　New Words and Phrases 121
    Part 3　Technical Words and Phrases 121
    Part 4　Translations 122
    Part 5　Reference Version 123

Section Ⅶ of Translating Skills：科普文章的翻译 125

Lesson 16 "The Sage of PID" ——An Educational Tale for Those Who Would Understand the Concept of PID(1) ……………… 128
   Part 1   Text ……………………………………………………………… 128
   Part 2   New Words and Phrases ……………………………………… 130
   Part 3   Reference Version ……………………………………………… 131
Lesson 17 "The Sage of PID"——An Educational Tale for Those Who Would Understand the Concept of PID(2) ……………… 133
   Part 1   Text ……………………………………………………………… 133
   Part 2   New Words and Phrases ……………………………………… 135
   Part 3   Reference Version ……………………………………………… 137

参考文献   **140**

Unit 1

Electromechanical Introduction

Lesson 1 • What Is Mechatronics?

Part 1 ▶▶ Text

In the mid-1980s, mechatronics came to mechanical engineering that is the boundary between mechanics and electronics. Today, the term encompassed a large array of technologies, many of which have become well-known in their own right. Each technology still has the basic element of the merging of mechanics and electronics but now may also involve much more particularly software and information technology. The relationship of mechatronics among the disciplines is illustrated in Fig. 1-1.

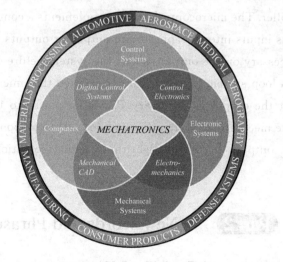

Fig. 1-1 The relationship of mechatronics among the disciplines

For example, many early robots resulted from mechatronics development. As robot systems become smarter, software development, in addition to the mechanical and electrical system, became central to mechatronics.

Mechatronics gained legitimacy in academic circles in 1996 with the publication of the first referred journal: IEEE/ASME Transactions on Mechatronics. In the premier issue, the authors worked to define mechatronics. After acknowledging that many definitions have circulated, they selected the following for articles to be included in transactions: "the synergistic integration of mechanical engineering with electronics and intelligent computer in the design and manufacturing of industrial products and processes." The authors suggested 11 topics that should fall, at least in part under the general category of mechatronics:

1. modeling and design
2. system integration
3. actuators and sensors
4. intelligent control
5. robotics
6. manufacturing
7. motion control
8. vibration and noise control
9. micro-device and optoelectronic systems
10. automotive
11. other application

A mechatronics design is a control system. One or more inputs are fed to a microcontroller. These inputs may have to undergo some signal conditioning before being read by the microcontroller. The microcontroller then implements a control algorithm that interprets the various inputs into the appropriate output or outputs. Again, signal conditioning may be necessary on the output side of the system before driving an actuator or display. In a closed loop system, feedback is received so that microcontroller is able to monitor and adjust the output as necessary. Providing power to the microcontroller is the last piece of the mechatronic system In summary, the components of a mechatronic system are input, output, a control algorithm, signal conditioning (if necessary), and power.

Part 2 New Words and Phrases

boundary ['baʊndri] n. 边界；范围；分界线
merge [mɜːdʒ] vt. 合并；使合并；吞没
discipline ['disiplin] n. 学科；纪律；训练；惩罚
robotic [rəʊ'bɒtɪk] n. 机器人学

microcontroller [ˌmaikrəukənˈtrəulə] n. 微控制器
implement [ˈɪmplɪment] vt. 实施，执行
　　　　　　　　　　　　n. 工具；手段
component [kəmˈpəunənt] n. 成分；组件；元件
in addition to　除…之外
in summary　总之；概括起来

Part 3 ▶▶ Technical Words and Phrases

electromechanical [ɪˌlektrəumɪˈkænɪkəl] adj. 电动机械的，机电的
mechatronics [mekəˈtrɒniks] n. 机电一体化；机械电子学
encompass [ɪnˈkʌmpəs] vt. 包含；包围，环绕；完成
legitimacy [liˈdʒitiməsɪ] n. 合法；合理；正统
micro-device [ˌmaikəudiˈvais] n. 微型装置；微型器件
automotive [ˌɔːtəˈməutɪv] adj. 汽车的；自动的
control algorithm　控制算法
mechatronic system　机电系统
mechanical engineering　机械工程；机械工程学
academic circle　学术界
motion control　运动控制，拖动控制；动作控制，移动控制
optoelectronic system　光电系统
IEEE　abbr. 电气与电子工程师协会（Institute of Electrical and Electronic Engineers）
ASME　abbr. 美国机械工程师协会（American Society of Mechanical Engineers）

Part 4 ▶▶ Translations

1. In the mid-1980s, mechatronics came to mechanical engineering that is the boundary between mechanics and electronics.

在20世纪80年代中期，机电一体化技术融入机械工程，它是介于机械工程与电子工程之间的交叉领域。

2. As robot systems become smarter, software development, in addition to the mechanical and electrical system, became central to mechatronics.

随着机器人系统的智能化发展，不仅是机械与电子系统，而且软件技术的发展也成为机电一体化的核心内容。

3. After acknowledging that, many definitions have circulated, they selected the

following for articles to be included in transactions: "the synergistic integration of mechanical engineering with electronics and intelligent computer in the design and manufacturing of industrial products and processes."

在列举了诸多公开的定义后,他们选择了如下内容发表在学报的文章上:"工业产品的设计与制造中机械工程、电子工程以及智能计算机的相互融合"。

4. A mechatronics design is a control system. One or more inputs are fed to a microcontroller.

机电一体化设计是一个控制系统。一个或多个输入信号被传输给一个微控制器。

5. The microcontroller then implements a control algorithm that interprets the various inputs into the appropriate output or outputs.

这些输入信号经过处理后,微控制器才能读取。然后微控制器进行控制运算,将各种输入信号转化为适当的输出信号。

6. In summary, the components of a mechatronic system are input, output, a control algorithm, signal conditioning (if necessary), and power.

总之,机电一体化系统是由输入、输出、控制运算、信号调节(若需要)以及动力源组成的。

Part 5　Reference Version

第一课　机电一体化简介

在20世纪80年代中期,机电一体化技术融入机械工程,它是介于机械工程与电子工程之间的交叉领域。今天,机电一体化包括了大量已经独立发展的技术,每一项技术仍然基于机械工程与电子工程的融合,同时又融入了很多新的技术,尤其是软件技术和信息技术。机电一体化技术与其他学科领域之间的关系如图1-1所示。例如:许多早期

图1-1　机电一体化技术与其他学科关系图

的机器人源于机电一体化的发展。随着机器人系统的智能化发展，不仅是机械与电子系统，而且软件技术的发展也成为机电一体化的核心内容。

随着 1996 年电气与 IEEE/ASME 的机电一体化学报的创刊，机电一体化技术在学术圈正式获得承认。在创刊号上，编者严格定义了机电一体化。在列举了诸多公开的定义后，他们选择了如下内容发表在学报的文章上："工业产品的设计与制造中机械工程、电子工程以及智能计算机的相互融合"。编者列举了 11 个方面属于或者至少是部分归类于机电一体化：

1. 建模与设计
2. 系统集成
3. 执行器与感应器
4. 智能控制
5. 机器人技术
6. 制造业
7. 运动控制
8. 振动与噪声控制
9. 卫星设备与光电系统
10. 汽车系统
11. 其他应用

机电一体化设计是一个控制系统。一个或多个输入信号被传输给一个微控制器。这些输入信号经过处理后，微控制器才能读取。然后微控制器进行控制运算，将各种输入信号转化为适当的输出信号。此外，在驱动执行器或者显示之前，信号波形加工对于系统的输出部分很必要。在一个闭环系统中，微控制器接收反馈后能根据需要检测和调整输出。在机电一体化系统中，为微处理器提供动力也是必不可少的。总而言之，机电一体化系统是由输入、输出、控制运算、信号调节（若需要）以及动力源组成的。

Lesson 2　Introduction of Modern Manufacturing Technology

 Part 1　Text

Over the years, manufacturing technology has achieved great progress with significant development of technology in other areas. Modern Manufacturing technology can be defined as the technology which is combined with mechanical, electronic, information and modern management technology applied in whole life cycle of product to achieve T (low lead time), Q (high quality), C (low cost) and S (good service) purposes. Following is not a complete list of all new technology that are used in manufacturing but some representative samples.

FMS stands for flexible manufacturing system. As the name suggests, the system can accommodate to certain changes in product types or the order of operations. FMS consists of three major categories: 1. computer controlled system; 2. automatic materials handling system; 3. numerically controlled machine tools. It can be considered to fill the gap between transfer lines which have low production rates but various types of parts. The relative position of FMS in manufacturing is illustrated in Fig. 2-1.

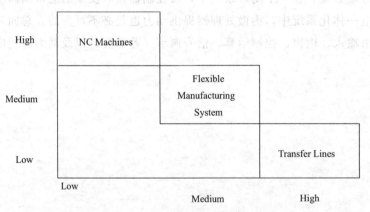

Fig. 2-1　The relative position of FMS in manufacturing

CIMS stands for computer integrated manufacturing system. It is a complex machine system based on the theory, and management in the whole manufacturing enterprise by computer integrated systems and data communication. If FMS is the manufacturing system on the shop level CIMS is the one on the enterprise level CIMS consists of several subsystems, such as CAD, CAPP, CAM, ERP (enterprise resource planning) and FMS.

RP is an acronym of rapid prototyping. Rapid prototyping 3D solids or surfaces model of particular part form certain CAD software, transforms them into cross sections, and then creates each cross section in physical space, one after the next until the model is finished. It is a WYSIWYG process where the virtual model and the physical model correspond almost identically. The first technologies for rapid prototyping began in the 1980s and were used to produce models and prototype parts. Today, they are used for a much wider range of applications. The typical and well developed RP technologies include: SLA (stereo-lithography apparatus), SLS (selective laser sintering), LOM (laminated object manufacturing) and FDM (fused deposition modeling).

VM means virtual manufacturing. It applied simulation, modeling and analysis technology in product development to predict capability and potential problems of manufacturability before real manufacturing. VM almost don't use any resources of energies compared to real manufacturing because it is just a virtual realization of product designing, developing and manufacturing in computer. The advantage of utilizing VM is fewer prototypes, less materials waste better quality and lower overall manufacturing cost.

Part 2 ▶▶ New Words and Phrases

define [di'fain]　　vt. 定义；使明确；
apply [ə'plai]　　vt. 申请；应用
Representative [ˌreprɪ'zentətɪv]　adj. 有代表性的
　　　　　　　　　　　　　　　　n. 代表；典型
flexible ['fleksəbl]　adj. 灵活的；柔韧的；易弯曲的
complex ['kɒmpleks]　adj. 复杂的；合成的
be combined with　　与…结合/联合
accommodate to　　适应
numerically [nju:'merɪklɪ]　adv. 数字上；用数字表示
subsystem ['sʌb'sɪstəm]　n. 子系统；次要系统

Part 3 ▶▶ Technical Words and Phrases

FMS　abbr. 柔性制造系统。能够在主体计算机的控制下，运用自动化装置从开始到完工生产产品系列，具有运用相同装配生产不同产品的能力。

CIMS　abbr. 计算机集成制造系统（Computer Integrated Manufacturing System）

CAD　*abbr*. 计算机辅助设计（Computer-Aided Design）

CAPP　*abbr*. 计算机辅助生产计划（Computer-Aided Process Planning）

CAM　*abbr*. 计算机辅助制造（computer-aided manufacturing）

　　　　　中央地址存储器（Central Address Memory）

ERP　*abbr*. 企业资源计划（Enterprise Resource Planning）

　　　　　有效辐射功率（effective radiated power）

RP　*abbr*. 菲律宾共和国（Republic of the Philippines）

　　　　　标准发音（Received Pronunciation）

　　　　　无线电传真（Radio-Photography）

　　　　　快速成型

transfer line　输送管路输电线，传输线

manufacturing technology　制造技术；制造工艺

computer integrated　计算机整合，计算机集成

Part 4　Translations

1. Modern Manufacturing technology can be defined as the technology which is combined with mechanical, electronic, information and modern management technology applied in whole life cycle of product to achieve T (low lead time), Q (high quality), C (low cost) and S (good service) purposes.

现代制造技术的定义为融合了机械、电子、信息和现代管理技术，应用于产品的整个生命周期，实现较少的研发时间（T）、高质量（Q）、低成本（C）和好的售后服务（S）目的的技术总称。

2. FMS consists of three major categories: 1. computer controlled system; 2. automatic materials handling system; 3. numerically controlled machine tools.

FMS 代表柔性制造系统。FMS 一般由三大类组成：1. 计算机控制系统；2. 自动物料传送系统；3. 数控机床。

3. It is a complex machine system based on the theory, and management in the whole manufacturing enterprise by computer integrated systems and data communication.

它是通过计算机集成系统和数据的通信将整个企业的人、技术和管理三者整合。

4. If FMS is the manufacturing system on the shop level, CIMS is the one on the enterprise level CIMS consists of several subsystems, such as CAD, CAPP, CAM, ERP (enterprise resource planning) and FMS.

如果 FMS 是车间级的制造系统，那 CIMS 就是公司级的。CIMS 有许多子系统，如 CAD \ CAPP \ CAM \ ERP（企业资源计划）和 FMS。

5. The first technologies for rapid prototyping began in the 1980s and were used to

produce models and prototype parts.

最早的快速成型技术始于 20 世纪 80 年代，用于生产模型和零件样品。

6. Today, they are used for a much wider range of applications. The typical and well-developed RP technologies include: SLA (stereo-lithography apparatus), SLS (selective laser sintering), LOM (laminated object manufacturing) and FDM (fused deposition modeling).

现在它被应用于更广泛的领域。典型的和成熟的 RP 技术包括：立体光固化成型、选择性激光烧结、分层实体制造和熔融沉积成型。

7. It applied simulation, modeling and analysis technology in product development to predict capability and potential problems of manufacturability before real manufacturing.

它将仿真、造型和分析技术应用于产品开发，以在真实制造前预知产品性能和制造工艺性的潜在问题。

8. The advantage of utilizing VM is fewer prototypes, less materials waste better quality and lower overall manufacturing cost.

使用虚拟制造技术的优点在于更少的样品制造、更少的材料消费、更高的质量和更低的整体制造成本。

Part 5 ▶▶ Reference Version

第二课 现代制造技术简介

这些年来，随着其他技术领域的显著发展，制造技术也取得了长足的进步。现代制造技术的定义为融合了机械、电子、信息和现代管理技术，应用于产品的整个生命周期实现较少的研发时间（T）、高质量（Q）、低成本（C）和好的售后服务（S）目的的技术总称。以下是应用于制造业的新技术的有代表性的例子，而非完整的列表。

FMS 代表柔性制造系统。顾名思义，系统能适应产品类型和工序顺序的某种改变。FMS 一般由三大类组成：1. 计算机控制系统；2. 自动物料传送系统；3. 数控机床。柔性制造系统被认为填补了生存率高但仅针对一种特定零件生产的传送线设备和生产率低但加工零件类型多的数控机床之间的空白。FMS 在制造业中的相对位置如图 2-1 所示。

CIMS 代表着计算机集成制造系统。它是基于 CIM 理论的复杂的人机系统。CIM 是一种新的管理哲学。它是通过计算机集成系统和数据的通信将整个企业的人、技术和管理三者整合。如果 FMS 是车间级的制造系统，那 CIMS 就是公司级的。CIMS 有许多子系统，如 CAD/CAPP/CAM/ERP（企业资源计划）和 FMS。

RP 是快速成型的首字母缩写。快速成型技术利用从某种 CAD 软件中产生的虚拟三维实体或表面模型，并将其转变成各个横截面，再在实际空间中逐个形成每一个横截

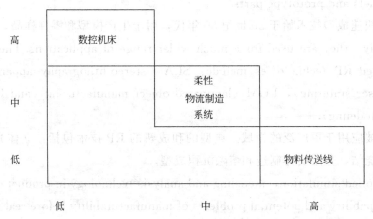

图 2-1　FMS 在现代制造业中的地位

面直至完成模型。这是个所见即所得的过程,虚拟模型和实际模型几乎完全一致。最早的快速成型技术始于 20 世纪 80 年代,用于生产模型和零件样品。现在它被应用于更广泛的领域。典型的和成熟的 RP 技术包括:SLA(立体光固化成型),SLS(选择性激光烧结),LOM(分层实体制造)和 FDM(熔融沉积成型)。

VM 即虚拟制造。它将仿真、造型和分析技术应用于产品开发,以在真实制造前预知产品性能和制造工艺性的潜在问题。虚拟制造和真实制造相比,无需使用任何资源和能量,因为它仅仅是产品设计、开发和制造在计算机中的虚拟再现而已。使用虚拟制造技术的优点在于更少的样品制造、更少的材料消费、更高的质量和更低的整体制造成本。

Section I of Translating Skills: 科技英语文体的特点与科技英语翻译概述

科技英语（English for Science and Technology，简称 EST）作为一种重要的英语文体，与非科技英语文体相比，具有词义多、长句多、被动句多、第三人称句多、词性转换多、非谓语动词多、专业性强等特点，这些特点都是由科技文献的内容所决定的。因此，科技英语的翻译也有别于其他英语文体的翻译。科技英语翻译必须遵循一定的翻译方法和翻译技巧。

科技英语翻译也有一定的标准，即准确规范、通顺易懂、简洁明晰。当然对译者也有一定的要求，它要求译者不仅要有一定的英语水平、汉语水平和专业知识，还要求译者掌握一定的翻译知识与技巧。常用的科技英语的一般翻译方法有：直译与意译；合译与分译；增译与省译；顺译与倒译。

第一节 翻译小练习

应用上面提到的翻译方法将下列英文句子翻译成汉语。

1. The power plant is the heart of a ship.
2. The power unit for driving the machines is a 50-hp induction motor.
3. Semiconductor devices, called transistors, are replacing tubes in many applications.
4. The designer must have access to stock lists of the materials he employs.
5. Part adjustment and repair must be performed on regular basis if an acceptable printed product is to be the end product.
6. These fragments of rock and iron range from thousand kilometers in diameter to less than one.
7. Manufacturing processes may be classified as unit production with small quantities being made and mass production with large numbers of identical parts being produced.
8. Cartography is the science of making maps.
9. The two units used most frequently in electricity are ampere and volt: this is the unit of voltage and that of current.
10. That like charges repel but opposite charges attract is one of the fundamental laws of electricity.
11. Almost any insulated body processes to some extent the ability to retain for a time an electric charge.

12. The angular contact bearing provides a greater thrust capacity.

13. The properties of the weld can be altered by varying the grain orientation.

第一节翻译小练习答案

1. 动力装置是船舶的心脏。

2. 驱动这些机器的动力装置是一台 50 马力的感应电动机。

3. 半导体装置也称为晶体管，在许多场合替代电子管。

4. 设计师必须备有所使用材料的储备表。（意译）

5. 要使印刷品的质量达到要求，部件调试及修理就必须定期进行。（意译）

6. 这些石块和铁块的碎片大小不等，大的直径有 1000 公里，小的不到 1 公里。（分译）

7. 制造过程可以分为单件生产和大量生产。前者指的是生产少量的零件，后者则是指生产大量相同的零件。（分译）

8. 制图学是研究绘制地图的科学。（增译）

9. 电学上最常用的两个单位是安培和伏特：后者是电压的单位，前者是电流的单位。（增译）

10. 同性电荷相斥，异性电荷相吸是电学的一个基本规律。（省译）

11. 几乎任何一种绝缘体都多少具有保留电荷一段时间的能力。

12. 向心推力轴承有较大的轴向承载能力。

13. 通过改变晶粒的方向可以改变焊缝的性能。

Unit 2

Electrical Technology

Lesson 3 • Direct Current and Alternating Current

Part 1 ▶▶ **Text**

Circuit is the passage of electrical current. It is to realize certain functions, which are assembled by electric apparatuses and components according to certain patterns. Power, inter-mediate connection and load are the components of circuit.

The actual circuits are of many types. Due to the different functions, they are classified into two types: electric power circuits and signal circuits. The electric power circuits is used for power transmission and transformation, such as electricity generation, power supply systems, electric drives and electric lighting. The electric signal circuits have many types, which have different functions. The production, amplification, shaping of various electric signals and calculation and storage of digital signals are all electric signal circuits.

A simple electrical circuit consists of a voltage source, a load, and connection lines. There are at least three electrical quantities in any circuit. These are current, voltage, and resistance. Voltage and current must be present in a circuit if electricity is to do work. Resistance is always present in every circuit because all conductors oppose the flow of electrons to some extent.

Physically, charge carriers-electrons-move through the electrical circuit via the electrical conductors from the negative pole of the voltage source to the positive pole. This motion of charge carriers is called electrical current. Current can only flow if the circuit is closed.

There are two types of current-direct current and alternating current as shown in Fig. 3-1:

• If the electromotive force in an electrical circuit is always in the same direction, the current also always flows in the same direction. This is called direct current (DC) or a DC circuit.

● In the case of alternating current or an AC circuit, the voltage and current change direction and strength in a certain cycle.

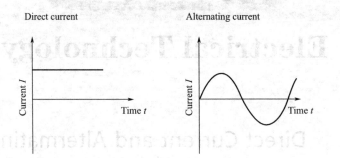

Fig. 3-1 Direct current and alternating current plotted against time

Fig. 3-2 shows a simple DC circuit consisting of a voltage source, electrical lines, a control switch, and a load (here a lamp).

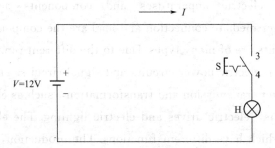

Fig. 3-2 DC circuit

When the control switch is closed, current I flows via the load. The electrons move from the negative pole to the positive pole of the voltage source. The direction of flow from quotes "positive" to "negative" was laid down before electrons were discovered. This definition is still used in practice today. It is called the technical direction of flow.

In an AC circuit, the sinusoidal alternating circuit is most commonly used. The simple sinusoidal alternating circuit refers to the circuit that contains sinusoidal power; besides, the voltage and current produced by each part of the circuit vary according to sine rules. The sinusoidal voltage may raise to lower the voltage with the transformer conveniently. So, the power station produces electric energy in simple sinusoidal alternating form. Then if utilizes the transformer to raise the voltage so as to transport the electricity over long distances. In a signal circuit, usually all kinds of sinusoidal signal

Unit 2
Electrical Technology

generators are used as signal sources. The voltage and current in a simple sinusoidal alternating circuit are sinusoidal quantities with the same frequency, which enables the quantities are also smooth. Generally, it is not likely to bring about too high a voltage so as to destroy the insulation of the electric installation. As a result, the sinusoidal quantity is widely used in electro-engineering.

Part 2 ▶▶ New Words and Phrases

passage ['pæsidʒ] *n.* 走廊；通路
inter-mediate [intə'mi:dieit] *n.* 中间物；媒介
classify ['klæsifai] *vt.* 分类；分等
amplification [ˌæmplifi'keiʃ(ə)n] *n.* 放大（率）；
calculation [kælkjʊ'leiʃ(ə)n] *n.* 计算；估计
oppose [ə'pəʊz] *vt.* 反对
motion ['məʊʃ(ə)n] *n.* 动作；移动
utilize ['ju:təlaiz] *vt.* 利用
frequency ['fri:kw(ə)nsi] *n.* 频率
insulation [insjʊ'leiʃ(ə)n] *n.* 绝缘；隔离
to some extent 在一定程度上
in practice 实际上

Part 3 ▶▶ Technical Words and Phrases

circuit ['sɜːkit] *n.* 电路
resistance [ri'zistəns] *n.* 电阻
electron [i'lektrɒn] *n.* 电子
quote [kwəʊt] *n.* 引用
sinusoidal [ˌsainə'sɔidl] *adj.* 正弦曲线的
sinusoidal alternating circuit 正弦交流电路
sine rule 正弦定理
sinusoidal quantity 简谐波
direct current 直流，直流电
interacting current 交流电流
electrical current 电流
electric apparatuses 电器；电气设备；电气装置
electric power circuit 电力线路

15

signal circuit　信号电路，信号线路
  electricity generation　发电
  power supply system　供电系统；建筑供配电系统
  electric drive　电力传动，电力驱动装置
  electric lighting　电气照明
  voltage source　电压电源
  charge carrier　电荷载子
  negative pole　阴极，负极
  positive pole　阳极；正极
  alternating current　交流电
  electromotive force　电动势
  lay down　放下；铺设

Part 4　Translations

  1. Circuit is the passage of electrical current. It is to realize certain functions, which are assembled by electric apparatuses and components according to certain patterns.

  电路是电流的通路，是为了实现某种功能，由电气设备和元器件按一定方式组合而成的。

  2. A simple electrical circuit consists of a voltage source, a load, and connection lines.

  简单的电路包括电压源、负载和连接线。

  3. Resistance is always present in every circuit because all conductors oppose the flow of electrons to some extent.

  由于所有导体多少都会对电子的流动产生抗性，电阻在电路中永恒存在。

  4. Physically, charge carriers-electrons-move through the electrical circuit via the electrical conductors from the negative pole of the voltage source to the positive pole.

  实际上，电荷载体（电子）是通过电子导体从电压源的正极向负极流动。

  5. If the electromotive force in an electrical circuit is always in the same direction, the current also always flows in the same direction. This is called direct current（DC）or a DC circuit.

  如果电路中的电动势总是在同一个方向，而电流也在同一方向流动，这种电流叫做直流电或直流线路。

  6. When the control switch is closed, current I flows via the load. The electrons move from the negative pole to the positive pole of the voltage source.

当控制开关关闭,电流 I 流经负载。电子从电压源的负极流向正极。

7. The simple sinusoidal alternating circuit refers to the circuit that contains sinusoidal power; besides, the voltage and current produced by each part of the circuit vary according to sine rules.

简单的正弦交流电路是指一种电路,它包含正弦电源、根据正弦定律产生的不同电压和电流。

8. The voltage and current in a simple sinusoidal alternating circuit are sinusoidal quantities with the same frequency, which enables the quantities are also smooth.

简单的正弦交流电路中的电压和电流与正弦量有着相同的频率,这样可以使正弦量平稳发出。

Part 5 Reference Version

第三课 直流电和交流电

电路是电流的通路,是为了实现某种功能,由电气设备和元器件按一定方式组合而成的。电源、中间环节和负载是电路的基本组成部分。

实际上电路种类繁多,按其功能的不同,可以分为两大类:电力电路和信号电路。电力电路主要用来实现电能的传输和转换,如发电、供电系统、电力拖动、电气照明等。信号电路主要用来实现信号的传递和处理。信号电路种类繁多、功能各异,各种电信号的产生、放大、整形,数字信号的运算、存储等都是信号电路。

简单的电路包括电压源、负载和连接线,且至少有三个因素存在,即电流、电压、电阻。

若电流在工作状态下,电压和电流必须存在。由于所有导体多少都会对电子的流动产生抗性,电阻在电路中永恒存在。

实际上,电荷载体(电子)是通过电子导体从电压源的正极向负极流动。电荷载体的这种运动被称为电流。电流只在线路闭合时流动。

有两种类型的电流:直流和交流,如图 3-1 所示:

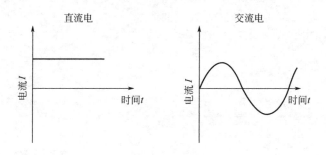

图 3-1 直流电流和交流电流随时间变化图

●如果电路中的电动势总是在同一个方向，而电流也在同一方向流动，这种电流叫做直流电或直流线路。

●交流电路中，电压和电流在一定周期内改变方向和强度。

如图 3-2 所示，简单的直流电路包括电压源、电线、控制开关和负载（这里是一盏灯）。当控制开关关闭，电流 I 流经负载。电子从电压源的负极流向正极。这是电流的专业流向。电子被发现之前，电流方向被认为是从正到负。现在人们仍旧这样认为。

在交流电路中，正弦交流电路最常用。简单的正弦交流电路是指一种电路，它包含正弦电源、根据正弦定律产生的不同电压和电流。正弦电压随着变压器相应地增加或减少。因此，发电站都是用正弦交流形式发电。如利用变压器提高电压来远距离传输电。在信号电路中，各种各样的正弦信号发生器通常作为信号源。简单的正弦交流电路中的电压和电流与正弦量有着相同的频率，这样可以使正弦量平稳发出。一般来说，它不太可能带来过高电压，破坏电气装置的绝缘性。因此，正弦量广泛用于电子工程行业。

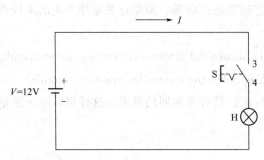

图 3-2 直流电路

Lesson 4 ● Resistor, Capacitor and Inductor

Part 1 ▶▶ Text

The circuits components adopted in various circuits mainly include two types: active components and passive components. Commonly used passive components control resistance units, inductance units and capacitance units. Commonly used active components contain voltage source and current supply.

The resistance component is the circuit component that reflects the physical phenomenon of current pyrometric effect. It is the component that consumes electric power and convents electric power into heat energy. In engineering, resistor is often used to realize current limiting, pressure release and partial pressure. The resistor and capacitor constitute filter and delayed relay, which may be used as decoupling resistance in power circuit. In all, resistance components have many effects in a circuit.

The inductance component is the circuit component reflecting the physical phenomenon of a magnetic field around current and it stored magnetic field energy. It is usually made of the hollow coils wounded by voltaic wires or coils with magnet core (shown in Fig. 4-1). The inductance component does not consume energy. It has the function of storing magnetic field energy. The magnetic filed energy stored by it at a certain time is directly proportional to the square of the current passing the component at that time. In engineering, the inductance component is normally used together with a capacitance component so as to form all types of high frequency and low frequency filter circuits,

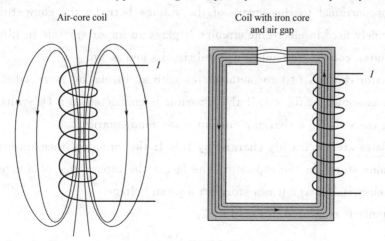

Fig. 4-1 Electrical coil and magnetic lines of force

frequency selection circuits, oscillation circuits, competition circuits and choked flow devices. It plays an important role in the circuit.

A magnetic field is induced when a current is passed through an electrical conductor. The strength of the magnetic field is proportional to the current. Magnetic fields attract iron, nickel and cobalt. The attraction increases with the strength of the magnetic field.

The solenoid has the following structure:

- The current-bearing conductor is wound around a coil. The overlapping of the lines of force of all loops increases the strength of the magnetic field resulting in a main direction of the field.

- An iron core is placed in the center. When current flows, the iron is also magnetized. This allows a significantly higher magnetic field to be induced with the same current (compared to an air-core coil).

These two measures ensure that a solenoid exerts a strong force on ferrous (= containing iron) materials.

In electropneumatic controls, solenoids are primarily used to control the switching of valves, relays or contactors. This can be demonstrated using the example of the spring-return directional control valve:

- If current flows through the solenoid coil, the piston of the valve is actuated.

- If the current is interrupted, a spring pushes the piston back into its initial position.

The capacitance competition is the circuit competition reflecting the physical phenomena the storage electrical charges produce electric field and the storage of electric field energy. The capacitance competition does not consume energy and has the function of storing electric field energy. Besides, the electric field energy stored at a certain time is directly proportional to the square of the voltage born by the competition at that time. It is widely used in electronic circuits. It plays an important role in filter circuits, turning circuits, coupling circuits, delayed circuits and so on.

A capacitor consists of two metal plates with an insulating layer (dielectric) between them as shown in fig. 4-1. If the capacitor is connected to a DC voltage source, then closing the switch, a charging current flows momentarily.

Both plates are electrically charged by this. If the circuit is then interrupted, the charge remains stored in the capacitor. The larger the capacitance of a capacitor, the greater the electrical charge it can store for a given voltage.

Capacitance is measured in Farad (F):

$$1F = 1\frac{As}{V}$$

If the charged capacitor is now connected to a load, then closing switch, the capacitor discharges. Current flows through the load until the capacitor are fully discharged.

Part 2 New Words and Phrases

constitute ['kɒnstɪtjuːt] vt. 组成，构成
coil [kɔɪl] vt. 盘绕
          n. 线圈
induce [ɪn'djuːs] vt. 诱导；引起
solenoid ['sɒlɪnɔɪd] n. 螺线管；螺线形电导管
exert [ɪg'zɜːt] vt. 运用，发挥
ferrous ['ferəs] adj. 亚铁的；铁的，含铁的
spring [sprɪŋ] n. 弹簧
          vt. 使弹开
momentarily ['məʊməntrəli] adv. 随时地，暂时地
be proportional to   与……成比例

Part 3 Technical Words and Phrases

resistor [rɪ'zɪstə(r)] n. 电阻器
capacitor [kə'pæsɪtə(r)] n. 电容器
Inductor [ɪn'dʌktə(r)] n. 感应器
filter ['fɪltə(r)] n. 滤波器；过滤器
capacitance [kə'pæsɪtəns] n. 电容，电流容量
circuit component   电路元件
active component   有效元件；作用分量
passive component   无源元件
resistance unit   电阻元件
inductance unit   电感元件；自感应元件
voltage source   电压电源
current supply   供电；电流源
resistance component   电阻元件
circuit component   电路元件
physical phenomenon   物理现象
current pyrometric effect   电流热效应
engineering resistor   工程电阻器

current limiting　电流限制
delayed relay　延时继电器
decoupling resistance　去耦电阻
inductance component　电感组件
magnetic field　磁场
hollow coil　空心线圈
voltaic wire　导线
magnet core　磁芯
capacitance component　电容组件
high frequency filter circuit　高频滤波电路
low frequency filter circuit　低频滤波电路
frequency selection circuit　选频电路
oscillation circuit　振荡电路；振荡回路
competition circuit　集成电路
choked flow device　阻流设备
electrical conductor　导电体，电导体；电导线

Part 4　Translations

1. The circuits components adopted in various circuits mainly include two types: active components and passive components.

各种电路中所采用的电路元件主要包括两种，分别为有源元件和无源元件。

2. Commonly used active components contain voltage source and current supply.

常用的有源元件包括电压源和电流源。

3. In all, resistance components have many effects in a circuit.

总之，电阻在电路中的作用很多。

4. It is usually made of the hollow coils wounded by voltaic wires or coils with magnet core.

一般由导线绕成空心线圈或带铁芯的线圈而成的。

5. The magnetic filed energy stored by it at a certain time is directly proportional to the square of the current passing the component at that time.

其在某时刻储存的磁场能量与该时刻流过元件的平方成正比。

6. In engineering, the inductance component is normally used together with a capacitance component so as to form all types of high frequency and low frequency filter circuits, frequency selection circuits, oscillation circuits, competition circuits and choked flow devices.

在工程中，电感通常和电容在一起使用，组成各种高低频滤波电路、选频电路、振荡电路、补偿电路及阻流器等，在电路中发挥着重要作用。

7. In electropneumatic controls, solenoids are primarily used to control the switching of valves, relays or contactors.

在电气控制方面，螺线形电导管主要是用来控制阀门的开关、继电器或接触器。

8. If current flows through the solenoid coil, the piston of the valve is actuated.

电流流经电磁线圈时，可以驱动阀门活塞。

9. A capacitor consists of two metal plates with an insulating layer (dielectric) between them.

电容器包括两个金属板，中间绝缘板（电介质）。

10. The larger the capacitance of a capacitor, the greater the electrical charge it can store for a given voltage.

电容器的电容越大，在一定电压下，可存储的电量越大。

Part 5 ▶▶ Reference Version

第四课　电阻器、电容器和电感器

各种电路中所采用的电路元件主要包括两种，分别为有源元件和无源元件。常用的无源元件包括电阻、电感、电容；常用的有源元件包括电压源和电流源。

电阻是反映电流热效应这一物理现象的电路元件，是一种消耗电能并转变为热能的元件。工程上常利用电阻来实现限流、降压、分压，其与电容器一起可以组成过滤器及延时继电器，在电源电路中可作为去耦电阻使用。总之，电阻在电路中的作用很多。

电感是反映电流周围存在磁场、储存磁场能量这一物理现象的电路元件，一般由导线绕成空心线圈或带铁芯的线圈而成的（见图 4-1）。电感不消耗能量，且具有储存磁场能量的功能，其在某时刻储存的磁场能量与该时刻流过元件的电流平方成正比。在工程中，电感通常和电容一起使用，组成各种高低频滤波电路、选频电路、振荡电路、补偿电路及阻流器等，在电路中发挥着重要作用。

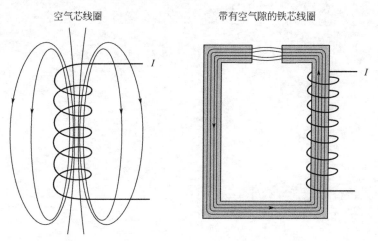

图 4-1　螺线形电导管（线圈）结构示意图

电流通过导体时可以感应到磁场。磁场强度与电流成正比。磁场吸引铁、镍和钴。磁场吸引力随着磁场的强度的增加而增加。

螺线形电导管结构：
- 载电导体由线圈缠绕。所有线圈重叠，增加磁场的强度，磁场运动方向一致。
- 中心为铁芯。电流开始，铁芯被磁化，产生更强大的磁场（与空心线圈相比）。

这两种模式会让螺线形电导管在铁质（含铁）材料上施加强力。

在电气控制方面，螺线形电导管主要是用来控制阀门的开关、继电器或接触器。回位弹簧换向阀可以证明这一点：
- 电流流经电磁线圈时，可以驱动阀门活塞。
- 电流中断，弹簧推动活塞回到其初始位置。

电容竞争就是电路竞争，反映了存储电荷产生电场和电场能量存储的真实现象。电容竞争不消耗能量，且具有存储电场能量的功能。而且，存储的电场能量与因竞争产生的电压成平方比。该定律在电路中应用广泛，在滤波电路、回转电路、耦合电路、延时电路等电路中均起着重要的作用。

电容器包括两个金属板，中间绝缘层（电介质），如图 4-1 所示。如果电容器连接到直流电压源，然后关闭开关，充电电流瞬间出现。

两个金属板都是通过这样的方式充电。如果电路中断，剩余电量仍在电容器中。电容器的电容越大，在一定电压下，可存储的电量越大。

电容的容量单位是法拉（F）：

$$1F = 1\frac{As}{V}$$

如果带电电容器连接到负载，然后关闭开关，电容器放电。电流流经负载，直到电容器没电。

Section II of Translating Skills：科技英语中词的翻译

正确选择词义是保证译文质量的中心问题。英汉两种语言在词义方面存在着很大的差异，一般来说，英语词义比较灵活多变，词的涵义范围比较宽，词义对上下文的依赖性比较大。汉语词义比较严谨精确，词语的涵义范围比较窄，词义的伸缩性和对上下文的依赖性比较小。而且英语中一词多用、一词多义、一词多类的现象相当普遍。英汉词义方面的这些差异，必然给翻译带来一定的困难。下面我们就从词义的选择、词义的引申、词类的转换、词的增译、词的省译等几个方面来阐述科技英语中词的翻译。

一、词义的选择

英汉两种语言都有一词多义的现象，但总的来说，在现代汉语中，同一个词的词义在不同的上下文中的差别比较小，一词多义的现象不如英语那么普遍。因此在翻译时要根据上下文来确定词的准确翻译。例如 wet 这个词，在不同的上下文中具有不同的意义。

1. He was wet to the skin.　他全身都湿透了。
2. Wet Paint.　油漆未干。
3. A wet country　允许贩酒的国家
4. wet behind the ears　乳臭未干的；没见过世面的

二、词义的引申

所谓词的引申，指的是在一个词所具有的基本词义的基础上，进一步加以引申，选择比较恰当的汉语词来表达，使原文的思想表现得更加准确，译文更加顺畅。如 heavy 的基本词义是"重"，但在下面这些情况中会引申成不同的含义。

　　heavy current　　强电流
　　heavy traffic　　交通拥挤
　　heavy industry　　重工业
　　heavy rain　　大雨
　　heavy frog　　浓雾

三、词性的转换

英语属于印欧语系，汉语属于汉藏语系，这两种不同语系的语言无论在词汇方面或在语法方面都有很大的不同。就词性来说，同一意思在不同语言中可以用不同词性来表达。因此，在英汉翻译的过程中不宜拘泥于原文的词性对号入座，而应根据汉语的行文习惯适当进行词性转换使译文通顺达意、自然流畅。

（一）形容词译成名词。科技英语往往习惯用表示特征的形容词及其比较级来说明物质的特性。因此，翻译时可以在这类形容词后加"度""性"等词使之成为名词。例如：

As most metals are malleable and ductile, they can be beaten into plates and drawn into wire.

由于大多数金属具有韧性和延展性，所以它们可以压成薄板和拉成细丝。

（二）动词译成名词。常规的词性转换还有动词转译成名词，这是因为英语中有一些动词的概念很难直接用汉语的动词来表达的缘故。例如：

It's well-known that neutrons act differently from protons.

大家知道，中子的作用与质子不同。

（三）名词译成形容词。英语译成汉语时，在忠实于原文的前提下，为使译文通顺、易懂，可以把名词译成形容词。例如：

In certain cases frictions is an absolute necessity.

在一定场合下摩擦是绝对必要的。

（四）名词、形容词和介词译成动词。英语中有许多含有动作意义的名词和由动词派生的名词，可根据汉语动词使用的灵活性和广泛性的特点，把名词、形容词和介词译成汉语动词。例如：

The acquaintance of science means mastering the law of nature.

认识科学意味着掌握自然规律。（名词译成动词）

It is possible to cut all thread forms and sizes on a lathe.

可以在车床上车削各种形状和尺寸的螺纹。（形容词译成动词）

In general, positive or negative rake tools can be used on stainlesss teel.

通常，正前角和负前角的刀具都可以用来加工不锈钢。（介词译成动词）

四、词的增译与词的省译

增词法与减词法就是在翻译时根据句法上、意义上或修辞上的需要增加或减少一些词，以便能更加忠实通顺地表达原文的思想内容。当然，增词与减词不是随意的，而是基于英汉两种语言表达方式的差异，适当地加入或减少（冠词、介词、代词、连词、动词等词的省译现象比较多见）一些词，从而使译文忠信流畅，这种情况在科技英语文献翻译时比较常见。例如：

These principles will be illustrated by the following **transition**.

这些原理将由下列演变**过程**来说明。（增词）

The temperature needed for this **processing** is lower than that needed to melt the metal.

这种加工**方法**所需的温度要低于熔化该金属的温度。（增词）

Different metals differ in *their* conductivity.

不同的金属具有不同的导电性。（减词）

第二节 翻译小练习

一、将下列句子翻译成汉语，注意句中某些词的恰当翻译。

1. The earth isn't completely round; it is slightly flattened at the poles.

2. Rate of penetration was found to be proportional to the net pressure applied by the tool.

3. Microcomputers have found their application in the production of genius sensors.

4. Plastics find wide application in our daily life.

5. Traditionally, NC programming has been performed offline with the machine commands being contained on a punched tape.

6. We should get familiar with different systems of units.

7. The metal casts well.

8. Safety in a power plant is of great importance.

9. Most modern transmitters employ solid state circuits.

10. Good lubrication keeps the bearings from being damaged.

11. This action externally appears like the discharge of a capacitor.

12. Atomic cells are small and very light, as compared to ordinary dry ones.

13. The largest and most expensive products cannot, because of their size, be testable in the factory.

14. The receiver will operate reliably no matter what happens.

15. The frequency, wave length and speed of sound are closely related.

16. Such an engine is called an internal combustion engine.

17. The fact that the concrete is slow in setting is no sign that it is of poor quality.

18. This equipment forms an integral part of many advanced systems.

19. Once inside the oven, panels are subjected to a temperature of 365°F.

20. Various precautions have been taken against leakage.

21. This new model should appeal to potential buyers.

22. These decision-making processes are applicable to the entire field of engineering design——not just to mechanical engineering design.

二、给下列专业术语选择正确的翻译。

（一）1. low damping _____ 2. low brake _____ 3. low limit _____
　　　4. low shot _____　　 5. low current _____ 6. low order _____
　　　7. low access _____   8. low key _____　　 9. low brass _____

a. 仰视拍摄　　b. 下限　　　c. 慢速存取　　d. 下半轴瓦　　e. 弱阻尼
f. 低位　　　　g. 低速制动器　h. 低强度电流　i. 低音调键

（二）1. solid angle　 2. solid bearing　 3. solid body　　4. solid line
　　　5. solid borer　 6. solid color　　 7. solid crankshaft　8. solid gold

9. solid measuer　　10. solid circuit　　　11. solid injection　　　12. solid lubricant

a. 固体　b. 钻头　c. 赤金　d. 立体角　e. 实线　f. 单色　g. 无气喷射　h. 体积
i. 整体轴承　　j. 实心曲轴　　k. 固体润滑剂　　l. 固态电路

第二节翻译小练习答案

一、将下列句子翻译成汉语，注意句中某些词的恰当翻译。

1. 地球并非完全是圆的，它的两极略扁平。（形容词）
2. 人们发现钻孔速度与工具所受的净压力成正比。
3. 微型计算机已经应用于智能传感器的生产中。
4. 塑料在我们的日常生活中得到广泛应用。
5. 通常，数控机床的编程是脱机完成的，指令载于穿孔带上。（技术性引申）
6. 我们应该熟悉各种（计量）单位制。
7. 这种金属有良好的铸造性能。
8. 安全在电厂是非常重要的。
9. 现代发报机大多采用固体电路。
10. 润滑良好可保护轴承不受损伤。
11. 这一作用从外表上看来像电容器的放电现象。
12. 和普通电池相比，原子电池体积小、重量轻。
13. 最大最贵的产品，因为体积大，在该工厂里是不能检验的。
14. 无论发生什么情况，这台接收机都将可靠地工作。
15. 频率、波长和声速三者是密切相关的。
16. 这种发动机成为内燃机。
17. 混凝土凝固得慢并不表示其质量差。
18. 这个设备是许多先进系统不可缺少的一部分。
19. 一旦进入烘干炉，板材就处于365华氏度的温度条件下。
20. 已采取了各种措施防止渗漏。
21. 这种新的模式，会吸引潜在的买家。
22. 这些判定过程可以应用于工程设计的整个领域，而不仅仅限于机械工程设计。

二、给下列专业术语选择正确的翻译。

（一）1. e　2. g　3. b　4. a　5. h　6. f　7. c　8. i　9. d

（二）1. d　2. i　3. a　4. e　5. b　6. f　7. j　8. c　9. h
　　　10. l　11. g　12. k

Lesson 5 ● Ohm's Law and Measurement in Electrical Circuits

Part 1 ▶▶ Text

Electrical current is the flow of charge carriers in one direction. A current can only flow in a material if a sufficient number of free electrons are available. Materials that meet this criterion are called electrical conductors. The metals copper, aluminum and silver are particularly good conductors. Copper is normally used for conductors in control technology.

Every material offers resistance to electrical current. This results when the free-moving electrons collide with the atoms of the conductor material, inhibiting their motion. Resistance is low in electrical conductors.

Materials with particularly high resistance are called insulators. Rubber and plastic-based materials are used for insulation of electrical wires and cables.

The negative pole of a voltage source has a surplus of electrons. The positive pole has a deficit. This difference results in source emf (electromotive force).

Ohm's law (Fig. 5-1) expresses the relationship between voltage, current and resistance. It states that in a circuit of given resistance, the current is proportional to the voltage, that is

- If the voltage increases, the current increases.
- If the voltage decreases, the current decreases.

$$V = R \cdot I \quad \begin{array}{ll} V = \text{Voltage}; & \text{Unit: Volt(V)} \\ R = \text{Resistance}; & \text{Unit: Ohm}(\Omega) \\ I = \text{Current}; & \text{Unit: Ampere(A)} \end{array}$$

Fig. 5-1 Ohm's law

In mechanics, power can be defined by means of work. The faster work is done, the greater the power needed. So power is "work divided by time". In the case of a load in an electrical circuit, electrical energy is converted into kinetic energy (for example electrical motor), light (electrical lamp), or heat energy (such as electrical heater, electrical lamp). The faster the energy is converted, the higher the electrical power. So here, too, power means converted energy divided by time. Power increases with current and voltage.

The electrical power of a load is also called its electrical power input (Fig. 5-2).

$$P = V \cdot I \quad \begin{array}{ll} P = \text{Power}; & \text{Unit: Watt(W)} \\ V = \text{Voltage}; & \text{Unit: Volt(V)} \\ I = \text{Current}; & \text{Unit: Ampere(A)} \end{array}$$

Fig. 5-2　Electrical power

Application example:

The solenoid coil of a pneumatic 5/2-way valve is supplied with 24 VDC. The resistance of the coil is 60 Ohm. What is the power?

The current is calculated by means of Ohm's law:

$$I = \frac{V}{R} = \frac{24\text{V}}{60\Omega} = 0.4\text{A}$$

The electrical power is the product of current and voltage:

$$P = V \cdot I = 24\text{V} \cdot 0.4\text{A} = 9.6\text{W}$$

Measurement means comparing an unknown variable (such as the length of a pneumatic cylinder) with a known variable (such as the scale of a measuring tape). A measuring device (such as a ruler) allows such measurements to be made. The result-the measured value-consists of a numeric value and a unit (such as 30.4 cm).

Electrical currents, voltages and resistances are normally measured with multimeters. These devices can be switched between various modes:

- DC current and voltage, AC current and voltage
- Current, voltage and resistance

The multimeter (Fig. 5-3) can only measure correctly if the correct mode is set. Devices for measuring voltage are also called voltmeters. Devices for measuring current are also called ammeters.

Before carrying out a measurement, ensure that voltage of the controller on which you are working does not exceed 24 V! Measurements on parts of a controller operating at higher voltages (such as 230 V) may only be carried out by persons with appropriate training or instruction. Incorrect measurement methods can result in danger to life.

Follow the following steps when making measurements of electrical circuits.

- Switch off voltage source of circuit.
- Set multimeter to desired mode. (voltmeter or ammeter, AC or DC, resistance)
- Check zeroing for pointer instruments. Adjust if necessary.
- When measuring DC voltage or current, check for correct polarity. ("+" probe of device to positive pole of voltage source).
- Select largest range.

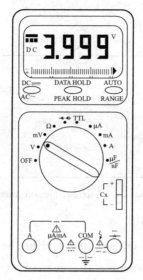

Fig. 5-3   Multimeter

- Switch on voltage source.
- Observe pointer or display and step down to smaller range.
- Record measurement for greatest pointer deflection (smallest measuring range).
- For pointer instruments, always view from vertically above display in order to avoid parallax error.

1) Voltage measurement (Fig. 5-4)

For voltage measurement, the measuring device (voltmeter) is connected in parallel to the load. The voltage drop across the load corresponds to the voltage drop across the measuring device. A voltmeter has an internal resistance. In order to avoid an inaccurate measurement, the current flowing through the voltmeter must be as small as possible, so the internal resistance of the voltmeter must be as high as possible.

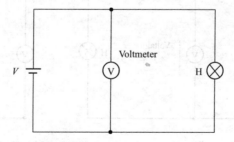

Fig. 5-4   Voltage measurement

2) Current measurement (Fig. 5-5)

For current measurement, the measuring device (ammeter) is connected in series to the load. The entire current flows through the device. Each ammeter has an internal resistance. In order to minimize the measuring error, the resistance of the ammeter must be as small as possible.

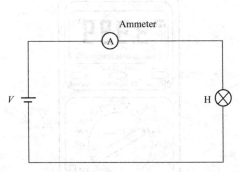

Fig. 5-5 Current measurement

3) Resistance measurement

The resistance of a load in a DC circuit can either be measured directly or indirectly.

• Indirect measurement measures the current through the load and the voltage across the load [Fig. 5-6(a)]. The two measurements can either be carried out simultaneously or one after the other. The resistance is then measured using Ohm's law.

• For direct measurement the load is separated from the rest of the circuit [Fig. 5-6(b)]. The measuring device (ohmmeter) is set to resistance measurement mode and connected to the terminals of the load. The value of the resistance is displayed.

If the load is defective (for example, the magnetic coil of a valve is burned out), the measurement of resistance either results in a value of zero (short-circuit) or an infinitely high value (open circuit).

Warning: The direct method must be used for measuring the resistance of a load in AC circuits.

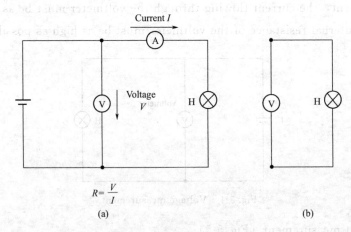

Fig. 5-6 Measuring resistance

Measuring devices cannot measure voltage, current and resistance to any desired degree of accuracy. The measuring device itself influences the circuit it is measuring, and no measuring device can display a value precisely. The permissible display error of a measuring device is given as a percentage of the upper limit of the effective range. For example, for a measuring device with an accuracy of 0.5, the display error must not exceed 0.5% of the upper limit of the effective range.

Application example (Table 5-1):

A Class 1.5 measuring device is used to measure the voltage of a 9 V battery. The range is set once to 10 V and once to 100 V. How large is the maximum permissible display error for the two effective ranges?

Table 5-1　Calculating the display error

Range	Permissible display error	Percentage error
10V	$10V \cdot \frac{15}{100} = 0.15V$	$\frac{0.15}{9V} \cdot 100\% = 1.66\%$
100V	$100V \cdot \frac{15}{100} = 1.5V$	$\frac{1.5}{9V} \cdot 100\% = 16.6\%$

The example shows clearly that the permissible error is less for the smaller range. Also, the device can be read more accurately. For this reason, you should always set the smallest possible range.

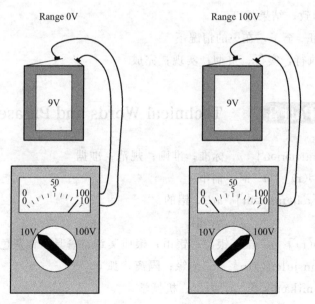

Fig. 5-7　Measuring battery voltage (with different range settings)

Part 2 ▶▶ New Words and Phrases

sufficient [səˈfɪʃnt]　*adj.* 足够的；充分的

inhibit [ɪnˈhɪbɪt]　*vt.* 抑制；禁止

insulator ['insjuleitə(r)] n. 绝缘体

deficit ['defisit] n. 赤字；不足额

convert [kən'vɜːt] vt. 使转变；转换

exceed [ɪk'siːd] vt. 超过；胜过

appropriate [ə'prəʊpriət] adj. 适当的；恰当的

probe [prəʊb] n. 探针；调查
vt. 探查

observe [əb'zɜːv] vt. 观察；遵守；注意到；评论

deflection [dɪ'flekʃ(ə)n] n. 偏向；挠曲；偏差

vertically ['vɜːtɪklɪ] adv. 垂直地

parallel ['pærəlel] n. 平行线；对比

correspond [ˌkɒrə'spɒnd] vi. 符合，一致；相应；通信

inaccurate [ɪn'ækjərət] adj. 错误的

minimize ['mɪnɪmaɪz] vt. 使减到最少

simultaneously [ˌsɪməl'teɪnɪəslɪ] adv. 同时地

defective [dɪ'fektɪv] adj. 有缺陷的；不完美的

collide with 冲突

result in 导致，结果是

in the case of 至于，在…的情况下

carry out 执行，实行；贯彻；实现；完成

Part 3 Technical Words and Phrases

criterion [kraɪ'tɪərɪən] n. 标准；准则；规范；准据

copper ['kɒpə(r)] n. 铜；铜币

aluminium [ˌæljə'mɪnɪəm] adj. 铝的
n. 铝

silver ['sɪlvə(r)] n. 银；银器；银币；银质奖章；餐具；银灰色

insulation [ˌɪnsjʊ'leɪʃ(ə)n] n. 绝缘；隔离；孤立

mechanics [mɪ'kænɪks] n. 力学；机械学

VDC abbr. 电压-数字变换器（Voltage-to-Digital Converter）

Ohm n. 欧姆（电阻单位）

multimeter ['mʌltɪmiːtə(r)] n. 万用表

voltmeter ['vəʊltmiːtə(r)] n. 伏特计，电压计

ammeter ['æmiːtə(r)] n. 安培计；电流计

zeroing ['zɪərəʊɪŋ] n. 零位调整
v. 做归零校正

polarity [pəˈlærəti] n. 极性；两极；对立
upper limit   上限；最高极限
free electron   自由电子
electrical conductor   导电体，电导体；电导线
free-moving electron   自由电子
plastic-based material   塑料为基础的材料
negative pole   阴极，负极
positive pole   阳极；正极
kinetic energy   动能
solenoid coil   电磁线圈，螺管线圈
pneumatic cylinder   气缸，气压缸
numeric value   数值，数字值
DC current and voltage   直流电流电压
AC current and voltage   交流电流电压
parallax error   视差；判读误差

Part 4 ▶▶ Translations

1. A current can only flow in a material if a sufficient number of free electrons are available.

电流只能在介质中流动，且自由电子的数量必须足够多。

2. Materials that meet this criterion are called electrical conductors. The metals copper, aluminum and silver are particularly good conductors.

可以让电流流动的介质即导体，如金属铜、铝、银。

3. This results when the free-moving electrons collide with the atoms of the conductor material, inhibiting their motion.

当自由运动的电子与导电材料的原子发生碰撞时，会导致阻碍这种运动。

4. The negative pole of a voltage source has a surplus of electrons. The positive pole has a deficit. This difference results in source emf（electromotive force）.

电源的负极处电子过剩，正极处不足，这种差别导致电动势。

5. Ohm's law expresses the relationship between voltage, current and resistance.

欧姆定律说明了电压、电流和电阻间的关系。

6. If the voltage increases, the current increases.

电压增加，电流增加。

7. So power is "work divided by time".

所以功率是单位时间内的功。

8. In the case of a load in an electrical circuit, electrical energy is converted into ki-

netic energy (for example electrical motor), light (electrical lamp), or heat energy (such as electrical heater, electrical lamp).

电路中若有负载,电能转化为动能(例如电机)、光(电灯)、或热能(如电加热器、电灯)。

9. Measurement means comparing an unknown variable (such as the length of a pneumatic cylinder) with a known variable (such as the scale of a measuring tape).

测量是为了某一未知变量(如气缸)和已知变量(如卷尺刻度)的比较。

10. The multimeter can only measure correctly if the correct mode is set.

万用表只能检测模式设置是否正确。

11. Follow the following steps when making measurements of electrical circuits.

测量电路时应遵循以下步骤:

12. For voltage measurement, the measuring device (voltmeter) is connected in parallel to the load.

测量设备(电压表)与载物并行连接。

13. In order to avoid an inaccurate measurement, the current flowing through the voltmeter must be as small as possible, so the internal resistance of the voltmeter must be as high as possible.

为了测量准确,通过电压表的电流必须尽可能小,因此电压表的内部阻力必须尽可能高。

14. For current measurement, the measuring device (ammeter) is connected in series to the load. The entire current flows through the device. Each ammeter has an internal resistance. In order to minimize the measuring error, the resistance of the ammeter must be as small as possible.

测量设备(电流表)与载物串联,确保全部电流通过。电流表有内部电阻。为减少测量误差,电流表的电阻必须尽可能小。

15. The resistance is then measured using Ohm's law.

电阻通过欧姆定律得出。

Part 5 ▶▶ Reference Version

第五课　欧姆定律和测量

电流是同一方向运动的电荷载体的流动。电流只能在介质中流动,且自由电子的数量必须足够多。可以让电流流动的介质即导体,如金属铜、铝、银。铜在控制技术中属常用导体。

物体具有阻碍电流流过的性质。自由电子和碰撞并抑制导体原子时介质会产生抗电流性质。电导体抗性低。

Unit 2
Electrical Technology

高抗性的材料被称为绝缘体。橡胶和塑料制品常常作为电线和电缆的绝缘体。

电源的负极处电子过剩，正极处不足，这种差别导致电动势。

欧姆定律（图 5-1）说明了电压、电流和电阻间的关系。在给定电阻的电路中，电流与电压成正比，即：

- 电压增加，电流增加。
- 电压降低，电流减少。

$$V=RI \quad \begin{aligned}&V\text{为电压，单位为伏特(V)}\\&R\text{为电阻，单位为欧姆}(\Omega)\\&I\text{为电流，单位为安培(A)}\end{aligned}$$

图 5-1 欧姆定律

从力学角度来看，通过做功可定义功率。做功越大，所需功率越大。所以功率是单位时间内所做的功。电路中若有负载，电能转化为动能（例如电机）、光（电灯）、或热能（如电加热器、电灯）。电能转换越快，功率越高。因此，功率即单位时间内转化能，意味着转换能量除以时间。功率与电流和电压成正比。

负载电功率即电功率输入（图 5-2）。

$$P=VI \quad \begin{aligned}&P\text{为功率，单位为瓦特(W)}\\&V\text{为电压，单位为伏特(V)}\\&I\text{为电流，单位为安培(A)}\end{aligned}$$

图 5-2 电功率

例如：

气动 5/2 换向阀的电磁线圈需要 24 伏直流电。线圈的电阻是 60 欧姆。功率是多少？

通过欧姆定律计算电流：

$$I=\frac{V}{R}=\frac{24\text{V}}{60\Omega}=0.4\text{A}$$

电流和电压算出功率：

$$P=V\times I=24\text{V}\times 0.4\text{A}=9.6\text{W}$$

测量是为了某一未知变量（如气缸的长度）和已知变量（如卷尺刻度）的比较。

测量装置（如尺子）允许这样的测量。结果（测量值）包含数值和单位（如 30.4 厘米）。

电流、电压和电阻通常用万用表测量，这类设备可以实现各种模式之间的切换：

- 直流电流和电压，交流电流和电压
- 电流、电压和电阻

万用表（图 5-3）只有当检测模式设置正确时才能够进行测量。测量电压的设备称为电压表。测量电流的称为电流表。

测量前，确保控制器的电压工作不超过 24V。控制器高压（如 230V）部分必须由

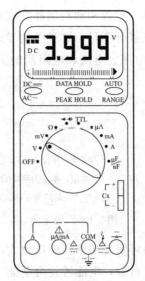

图 5-3 万用表

经过培训或指导的人员进行测量。不正确的测量方式会危及生命。

测量电路时应遵循以下步骤：
- 关闭电源。
- 调整万用表为所需的模式。（电压表、电流表、交流或直流、电阻）。
- 检查指针归零，如需要适当调整。
- 测量直流电压或电流时，检查极性。（"＋"探测正极）。
- 选择最大范围。
- 打开电源。
- 观察指针或显示器，并逐渐调到稍小范围。
- 记录指针最大偏差（最小测量范围）。
- 指针式仪器应从上方垂直观看，以避免视差。

1）电压测量（图 5-4）

测量设备（电压表）与负载并联连接。通过目标，电压降低，相当于电压在通过测量设备时降低。电压表有内部电阻。为了测量准确，通过电压表的电流必须尽可能小，因此电压表的内部阻力必须尽可能高。

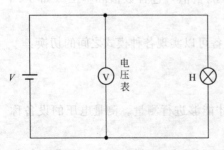

图 5-4 电压测量方法

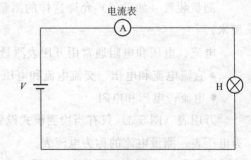

图 5-5 电流测量方法

2) 电流测量（图 5-5）

测量设备（电流表）与负载串联，确保全部电流通过。电流表有内部电阻。为减少测量误差，电流表的电阻必须尽可能小。

3) 电阻测量

直流电路中负载的电阻可以直接或间接测量。

● 间接测量方式用测量通过负载的电流和负载两端电压进行［图 5-6(a)］。两种测量可以同时或逐一测量。电阻通过欧姆定律得出。

● 直接测量方式中负载与电路的其余部分分开［图 5-6(b)］。测量设备（欧姆计）设定为电阻测量模式，连接在负载的接线端子两端，即可得电阻值。

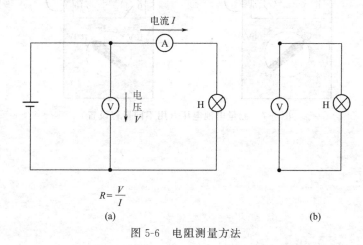

图 5-6　电阻测量方法

如果负载有故障（如电磁线圈的阀被烧毁），电阻的测量结果为零（短路）或无穷大（开路）。

警告：直接测量方式必须用于测量交流电路中负载的电阻。

测量设备不能精确地测出电压、电流、电阻。测量设备本身也会影响测量结果，没有测量设备可以测出相当精准的结果。在有效量程内，测量设备出现百分之几的误差都是可接受的。如：测量设备的精确度为 0.5，显示误差不得在有效范围超过 0.5% 的上限。

例如（表 5-1）：

1.5 级的测量设备用于测量 9V 电池的电压，量程是一次 10V，一次 100V，最大有效误差范围是多少？

示例清楚地表明：小量程有效误差较小，同时，可以较准确地读取结果。因此，应该设置最小量程。

表 5-1　计算显示误着

量程	允许显示误差	误差百分比
10V	$10V \cdot \dfrac{15}{100} = 0.15V$	$\dfrac{0.15}{9V} \cdot 100\% = 1.66\%$
100V	$100V \cdot \dfrac{15}{100} = 1.5V$	$\dfrac{1.5}{9V} \cdot 100\% = 16.6\%$

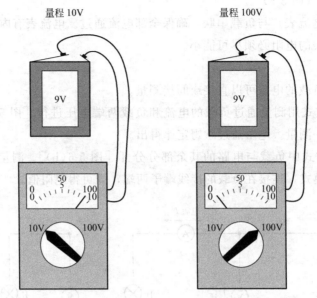

图 5-7 测量电池电压（用不同量程设置）

Unit 3

Sensor Detection Techniques

Lesson 6 • Introduction of Sensor Technology

 Part 1   ▶▶ Text

The words "sensor" and "transducer" are both widely used in the description of measurement systems. The former is popular in the USA whereas the latter has been used in Europe for many years. A dictionary definition of "sensor" is a device that detects a change in a physical stimulus and turns it into a signal which can be measured or recorded; a corresponding definition of "transducer" is "a device that transfers power from one system to another in the same or in the different form".

A sensible distinction is to use "sensor" for the sensing element itself and "transducer" for the sensing element plus any associated circuitry. All transducers would thus contain a sensor and most (though not all) sensors would also be transducers.

Fig. 6-1 shows the sensing process in terms of energy conversion. The form of the output signal will often be a voltage analogous to the input signal, though sometimes it may be a wave form whose frequency is proportional to the input or a pulse train containing the information in some other form.

Fig. 6-1 The sensing process

Sensors are particularly useful for making in situ measurements such as in industrial process control. Sensors are an important part to any measurement and automation application. The sensor is responsible for converting some type of physical phenomenon into a quantity measurable by a data acquisition (DAQ) system. Sensors are used in our everyday life in objects like elevator buttons, touch lamps, and many others. You see, in laymen's terms, sensors are devices that respond when a touch signal is received.

When we choose a kind of sensors, some factors could be to consider.

● Accuracy-The statistical variance about the exact reading.

● Calibration-Required for most measuring systems since their readings will drift over time.

● Cost.

● Environmental-Sensors typically have temperature and/or humidity limits.

● Range-Limits of measurement or the sensor.

● Repeatability-The variance in a sensor's reading when a single condition is repeatedly measured.

● Resolution-The smallest increment the sensor can detect.

There are two main types of sensors, the optical sensor and the microwave sensors. The first of the two main types of sensors are the optical sensors. As the name implies, these are sensors that, well, react to light. When visible lights are present, these sensors will react. Among the many kinds of light, these sensors use infrared lights and rays to activate them.

There are two subcategories of observation methods where these types of sensors are used. The first are the visible infrared remote sensing. This is the method wherein visible light are reflected by the objects on ground. It works by enabling the examination of the reflection strength as well as the land surface condition to understand it better. The downside of these types of sensors is that it cannot be used when there is darkness. Also, you cannot rely on these methods when it is cloudy or when the sunlight isn't too bright.

The second subcategory would be the thermal infrared sensing. You see, from the name itself, you will be able to understand that the sensors are picking up the signals through heat. When the infrared rays are emitted from the land surface, these sensors will get them and are able to show images basing on the radiation strength. This is a good method to use when the first one is not available.

The other half of the types of sensors is microwave sensors. These are sensors that, well, utilize the use of microwave signals. These are mainly used when the visible light as well as infrared rays are not available anymore. These are also used when the observation methods are not affected by weather or light. When it comes to microwave sensors, there are two observation types as well passive and active.

The types of sensors used in the active type observation method works by emitting microwave signals that are reflected in the land surface. This allows the observation of the land bodies such as valleys and mountains to be efficient and accurate. The second observation type, passive, uses the types of sensors by observing the microwave signals directly radiated from the sea surface. This therefore is used mostly in observing the temperature of sea surface as well as the ice thickness of a certain area. This type of observation methods are great because they are not affected by weather and air moisture, hence allowing a more accurate observation and more detailed results when it comes to the land details.

Always keep in mind that the wavelengths that are measured by both types of sensors are dependent to them. Getting to know these two different types of sensors will allow you to be able to choose the right types of sensors to use at certain observation scenarios.

Part 2　New Words and Phrases

description [dɪˈskrɪpʃ(ə)n]　*n.* 描述，描写；类型；说明书

whereas [weərˈæz]　*conj.* 然而；鉴于；反之

device [dɪˈvaɪs]　*n.* 装置；策略；图案

detect [dɪˈtekt]　*vt.* 察觉；发现；探测

signal [ˈsɪɡn(ə)l]　*n.* 信号；暗号；导火线

corresponding [ˌkɒrɪˈspɒndɪŋ]　*adj.* 相当的，相应的；一致的；通信的
　　　　　　　　　　　　　　　　v. 类似（correspond 的 ing 形式）；相配

transfer [trænsˈfɜː；traːns-；-nz-]　*n.* 转让；转移；传递；过户
　　　　　　　　　　　　　　　vt. 使转移；调任
　　　　　　　　　　　　　　　vi. 转让；转学；换车

sensible [ˈsensɪb(ə)l]　*adj.* 明智的；明显的；意识到的；通晓事理的

distinction [dɪˈstɪŋkʃ(ə)n]　*n.* 区别；差别；特性

associated [əˈsəʊʃɪ, eɪtɪd]　*v.* 联系（associate 的过去式和过去分词）
　　　　　　　　　　　　　　adj. 关联的；联合的；相关的 相应的 联合的

circuitry [ˈsɜːkɪtrɪ]　*n.* 电路；电路系统；电路学；一环路

conversion [kənˈvɜːʃ(ə)n]　*n.* 转换；变换

wave [weɪv]　*n.* 波动；波浪；高潮；挥手示意；卷曲

frequency [ˈfriːkw(ə)nsɪ]　*n.* 频率；频繁

proportional [prəˈpɔː(ʃ)n(ə)l]　*n.* [数] 比例项
　　　　　　　　　　　　　　　adj. 比例的，成比例的；相称的，均衡的

laymen [ˈleɪmən]　*n.* 非专业人员（外行）

calibration [ˌkælɪˈbreɪʃ(ə)n]　*n.* 校准；刻度；标度

optical [ˈɒptɪk(ə)l]　*adj.* 光学的；眼睛的，视觉的

subcategories　*n.* 子分类；子范畴（subcategory 的复数）

visible ['vɪzəbl] adj. 明显的；看得见的；现有的；可得到的
infrared [ɪnfrə'red] n. 红外线
　　　　　　　　　　adj. 红外线的　infrared rays 红外线
images ['ɪmɪdʒ] n. 图像，影像，肖像，想象
radiation [reɪdɪ'eɪʃ(ə)n] n. 辐射；发光；放射物
utilize ['juːtɪlaɪz] vt. 利用，利用 运用 使用
available [ə'veɪləb(ə)l] adj. 有效的，可得的；可利用的；空闲的，可用，可得到的

Part 3 ▶▶ Technical Words and Phrases

sensor ['sensə] n. 传感器
transducer [trænz'djuːsə；traːnz-；-ns-] n. 传感器
measurement systems　测量系统
physical stimulus　物理刺激
sensing element　[电子] 敏感元件；[自] 传感元件
voltage ['vəʊltɪdʒ；'vɒltɪdʒ] n. [电] 电压
proportional [prə'pɔːʃ(ə)n(ə)l] n. [数] 比例项
pulse [pʌls] n. [电子] 脉冲；脉搏
pulse train　脉冲序列；脉冲群；一串脉冲
data acquisition（DAQ）数据采集（DAQ）是指测量：电压、电流、温度、压力、声音、编码数据等电气或物理现象的过程
statistical [stə'tɪstɪk(ə)l] adj. 统计的；统计学的
variance ['veərɪəns] n. [数] 方差

Part 4 ▶▶ Translations

1. The latest Japanese vacuum cleaners contain sensors that detect the amount of dust and type of floor.

日本最新款吸尘器装有传感器，能测出灰尘量和地板类型。

2. A successful measurement system must be flexible enough to satisfy the information needs of an organization.

一个成功的度量系统必须是足够灵活的以满足组织的信息要求。

3. Piezoelectric ceramic is used here as the sensing element.

该传感器采用压电陶瓷作为敏感元件。

4. The switch works by passing a pulse of current between the tip and the surface.

电源开关的工作原理是让电流脉冲穿过触点和触面之间。

5. Designing and implementing data acquisition.

跟读设计并实现数据采集。

6. Digital RadioFrequency Memory (DRFM) is a kind of microwave signal storage device.

数字射频存储器（DRFM）是一种微波信号存储部件。

Part 5　Reference Version

"传感器"和"变送器"都是广泛用于描述测量系统的两个词语。前者在美国应用广泛，然而后者在欧洲已经使用了很多年。字典中对"传感器"作了定义，它是一个监测装置，用于探测某种物理反应并将这种反应转换成一个可以被测量和记录的信号；相应地也给出了"变送器"的定义，它是一个可以以相同或不同的形式将能量从一个系统传递给另一个系统的装置。

两者明显的区别就在于使用"传感器"应用的是传感元件本身，而使用"变送器"应用的是传感元件和任意相关的电路。所有变送器将包含一个传感器而大部分（但不是全部）的传感器都可以叫做变送器。

图 6-1 显示了能量转换的传感过程。输出信号的形式往往会是一个类似于输入信号的电压，尽管有时它可能是一个输入频率成正比的波形或是一个包含一些其他形式信息的脉冲序列。

输入能量（或信号）→变送器→输出能量（或信号）

图 6-1　传感过程

对于进行现场测量，传感器特别有用。传感器是任何测量和自动化应用中一个最重要的部分。传感器的主要功能是将某些类型的物理现象转换成由数据采集测量（DAQ）系统采集数据。在我们生活中用到传感器装置，比如电梯按键触屏，触摸灯，还有其他材料。要知道，非专业人员认为传感器就是当接收到一个触摸信号并作出反应的装置。

当我们选择传感器时，应考虑以下这些因素：

- 准确性——关于精确读数的统计方差。
- 刻度定位——读数将随时间而变化，大多数测量系统都需要刻度定位。
- 成本。
- 环保——传感器通常有温度或湿度的限定。
- 范围——测量和传感器的限制。
- 再现性——当同一个条件经过多次测量，传感器的读数将发生变化。
- 分辨率——传感器可以探测最小的增量。

传感器主要有光传感器和微波传感器两种。其中光传感器使用较广泛。顾名思义，这些传感器能够对感受到光线做出反应。当有一些可见光出现时，这些传感器将做出反应。这些传感器还可以使用红外线和射线来激活各种各样的光。

在使用这些类型的传感器时，有两个子分类的观测方法。第一个子分类观测方法是可见的红外遥感。该方法是指可见光反射到地面的物体上。通过反射强度以及地表条件去决定是否能更好地反馈信号。这两种传感器的缺点在光线阴暗的环境下均不可使用。另外，当遇到多云或是阳光不是特别明媚的时候，不能完全依靠这些方法。

　　第二个子分类是热红外遥感。从名称上看，你会认为这类传感器是通过热源来收集信号。当地表有红外线射出时，这些传感器就会接收红外线并且根据红外线的辐射强度显示图像。

　　另一类是微波传感器。这些传感器，利用对微波信号的使用。当无法检测到可见光和红外线时，会使用微波传感器。当我们使用微波传感器时，观测方法不会受天气和光线的影响。微波传感器的观测方法包括被动观测法和主动观测法。

Section Ⅲ of Translating Skills：科技英语中长句的翻译

科技英语中长句的比例很大，长句的一大特点是修饰语较长，一般为短语和从句，或是位于动词后面的状语短语或从句，这些修饰语中还可以一个套一个，甚至一连套几个。因此，长句的翻译是一个比较复杂、棘手的难题。翻译长句时，首先要弄清原文的句法结构：先找出句中的主要成分，即主语和谓语动词；再找出宾语、状语、定语等。然后分析出成分之间的逻辑关系，再按照汉语习惯正确地译出原文的意思。长句常见的翻译方法主要有以下四种：

➢ 化整为零，分译法；
➢ 逆序而下，顺序法；
➢ 逆流而上，逆序法；
➢ 纲举目张，变序法。

一、化整为零，分译法

有时英语长句中主句与从句或主句与修饰语间的关系不十分密切，翻译时可按照汉语多用短句的习惯，把长句中的从句或短语化为句子，分开来叙述，将原句化整为零。为使译文通顺连贯，也可以适当加几个连接词，就成了一个个单独的分句。例如：

So physics in the 19th century appeared to be devided into a few sciences or branches: mechanics, heat, sound, optics, and electro magnetism, with little or no connection between them.

因此，物理学在十九世纪就分成了几门学科或分支：力学、热学、声学、光学和电磁学。但是在这些学科之间很少或没有什么联系。

二、逆序而下，顺序法

当叙述的事物是按时间或逻辑顺序排列，与汉语表达方式基本一致时，翻译时常采用顺译法。例如：

(1) For these forms of pollution as for all the others, (2) the destructive chain of cause and effect goes back to a prime cause: (3) too many cars, too many factories, more and more trials left by supersonic jutes, inadequate methods for disinfecting sewers, too little water, too much carbon monoxide.

这个句子是由一个主句、一个状语和一个同位语组成的。主句是（2），破坏的因果关系链可归根于一个主要原因，也是全句的中心内容。主句前的（1）是方式状语，（3）是主句的同位语。

原文各句的逻辑关系、表达顺序与汉语完全一致，因此可以按照原文的顺序翻译。

译文：这些形式的污染像所有其他形式的污染一样，起破坏性的因果关系链可归根于一个主要的原因：太多的汽车、工厂、洗涤剂、杀虫剂，越来越多的喷气式飞机留下的尾气，不足的污水消毒处理方法，太少的水源，太多的一氧化碳。

三、逆流而上，逆序法

当英语长句的顺序与汉语表达方式不一致时，翻译时常常使用转换、颠倒、改变部分或完全改变词序的逆译法。例如：

1. (1) There is an equilibrium between the liquid and its vapor, (2) as many molecules being lost from the surface of the liquid and (3) then existing as vapor, (4) as reenter the liquid in a given time.

这个句子是由一个主句(1)，定语从句(4)以及两个分次独立结构组成，这个句子分为四层意思。(1)液体和固体之间处于平衡状态，是主句，是全句的中心所在；(2)许多分子从液体表面逸出，成为蒸汽；(3)又有同样多的分子重新进入液体；(4)在一定时间内。

根据汉语表达习惯，"因"在前面，"果"在后，可逆原文翻译。

译文：在一定时间内，许多分子从液体表面逸出，成为蒸汽。又有同样多的分子重新进入液体，因次，在液体和蒸汽之间处于平衡。

2. On account of the fact that there is always a resistance due to friction whenever one part of a machine moves over another, some work must be done in moving the parts of the machine itself.

根据汉语表达习惯，这句话同样用逆译法翻译比较贴切。

译文：机器的某个部件在另一个部件上面移动，总会有摩擦而产生阻力，由于这种情况，机器本身的部件运动时就要做一些功。

四、纲举目张，变序法

有时为了翻译的需要，我们要改变句子的顺序进行翻译，尤其是在翻译定语从句的句子时。例如：

The loads a structure is subjected to are divided into dead loads, which include the weights of all the parts of the structure, and live loads, which are due to the weights of people, movable equipment, etc.

［初译］ 一个结构物受到的荷载可分为包括结构物各部分重量的静载和由于人及可移动设备等的重量引起的活载。

［改译］ 一个结构物受到的荷载可分为静载与活载两类。静载包括该结构物各部分的重量。活载则是由于人及可移动设备等的重量而引起的荷载。

第三节 翻译小练习

认真分析下列句子的结构，并将它们翻译成汉语。

1. Gears play such an important part in machines that they have become the symbol

for machinery.

2. I am going to start up the generator in case the power goes off.

3. Electricity flows through a wire just as water flows through a pipe.

4. Even if someone is not a programmer, he can have a great impact on piece of software by suggesting how to improve it to the development team.

5. It is anticipated that LSI chips will show twice as high a reliability as they have now.

6. It is very interesting to note the differently chosen operating mechanism by the different manufacturers, in spite of the fact that the operating mechanism has a major influence on the reliability of the circuit-breakers.

7. These motors are, as a rule, especially designed for the specific machines with which they are used.

8. A machine spindle has different speeds which the operator selects according to the work.

9. Since an ammeter must be placed in series with the circuit, the latter must be opened for insertion of the meter.

10. Television equipment makes use of devices that can be considered optical because they depend on light rays for their operation.

11. The adjustment must be repeated each time the vessel is changed.

12. Each time the short circuit happens, the short circuit current runs the battery down.

13. In general, there is a measurable difference in potential between the two electrodes whether the cell is passing a current or not.

14. Deformation is affected by the stress inherent in the metal, the microstructural characteristics of the starting material, the temperature at which the deformation occurs, the rate at which the deformation occurs, and the frictional restraint between the material being forged and the die surface.

15. It is very important that a machine element be made of a material that has properties suitable for the condition of service as it is for the loads and stresses to accurately determined.

16. Although there is reason to hope that superconductivity may one day be found to exist in some materials at room temperature, for the moment is a phenomenon of the utterly cold.

17. Moving around the nucleus are extremely tiny particles, called electrons, which revolve around the nucleus in much the same way as the nine planets do around the sun.

第三节翻译小练习答案

1. 齿轮在机器中起着相当重要的作用，以致它已成为机械的象征。
2. 我去启动发电机，以免断电。
3. 电流通过导线，正像水流通过管道一样。
4. 即使一个人不是专业编程人员，他也可以通过向开发小组提出改进建议，而对一个软件产生极大的影响。
5. 预计大规模集成电路芯片的可靠性比目前可提高 1 倍。
6. 尽管操作装置对断路器的可靠性具有主要的影响，但注意不同的制造厂按不同形式选择操作装置是非常有趣的。
7. 总的来说，这些电动机是为使用它们的专门机器特别设计的。
8. 机床主轴有几种不同的速度，操作人员可以根据工件进行选用。
9. 由于电流计必须与电路串联，故后者必须断开以插入电流计。
10. 电视装置所用的器件可看成光学器件，因为这些器件是靠光线来工作的。
11. 每当更换容器时都要重新调整。
12. 每当短路发生时，短路电流就将电池的电能耗尽。
13. 通常，不论电池有无电流流过，两极之间均具有可测的电位差。
14. 变形会受到金属的内在应力、原始材料的微观结构特征、变形时的温度、变形的速度以及被锻造的材料与模具表面间的摩擦阻力的影响。
15. 载荷和应力必须准确地计算出来，机器零件要用性能符合工作条件的材料来制造。这两件事都是非常重要的。
16. 虽然有理由指望有一天可以发现某些材料在室温下会具有超导性，但目前超导性还只是一种极不热门的现象。
17. 围绕着原子核运动的是一些极其微小的粒子，称为电子，这些电子围绕着原子核旋转，正像九大行星围绕着太阳旋转一样。

Lesson 7 ◦ Capacitive and Inductive Proximity Sensors

Part 1 ▶▶ **Text**

The operational principle of a capacitive proximity sensor (Fig. 7-1) is based on the evaluation of the change in capacitance of a capacitor in an RC resonant circuit. The capacitance increases, when an object approaches the proximity sensor. This leads to a change in the oscillating action of the RC circuit which can be evaluated. The change in capacitance largely depends on the distance, the dimensions and the dielectric constant of the respective material.

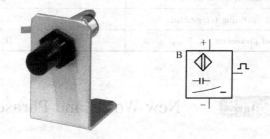

Fig. 7-1 Capacitive proximity sensors

Capacitive proximity sensor can detect metal objects, and can detect non-metallic objects, the metal objects can obtain the maximum distance, the action distance of the non metallic objects such as plastic, glass, water, oil, etc. is determined by the dielectric constant.

The proximity sensor has a PNP output, i. e. the signal line is switched to positive potential in the switched status. The switch is designed in the form of a normally open contact.

The load is connected between the sensor signal output and earth. A yellow light emitting diode (LED) indicates the switching status.

The capacitive proximity sensor cannot be flush fitted.

During operation, please observe the polarity of the applied voltage. The terminals are colour coded (shown in Table 7-1).

Table 7-1 Terminal colour code

Parameter	Value
Positive terminal	brown
Negative terminal	blue
Load output	black

Technical data for capacitive proximity sensors are shown in Table 7-2.

Table 7-2 Technical Data

Parameter	Value
Permissible operating voltage	10...55 VDC
Switch output	PNP, Normally open contact
Nominal switching distance	2...8mm
Hysteresis(at nominal switching distance)	3...15%
Maximum switching current	200mA
Maximum switching frequency	300Hz
Current consumption during idling(at 55 V)	7mA
Permissible ambient operating temperature	20℃...+70℃
Degree of protection	IP 65

Part 2 New Words and Phrases

capacitive [kəˈpæsɪtɪv] *adj.* 电容性的
proximity [prɒkˈsɪməti] *n.* 接近，亲近
capacitance [kəˈpæsɪtəns] *n.* 电容；电流容量
resonant [ˈrezənənt] *adj.* 共鸣的，洪亮的，反响的；谐振
polarity [pəˈlærəti] *n.* 有两极，极性，磁性引力

Part 3 Technical Words and Phrases

dielectric constant 介电常数
positive potential 正电势，正电位
normally open contact 常开触点
light-emitting diode 发光二极管
earth 大地，地；接地；地线；搭铁

【注】什么是介电常数？
介电常数：英文名称 dielectric constant。电容器的极板间充满电介质时的电容与极

板间为真空时的电容之比值称为（相对）介电常数，又称为"电容率"或"相对电容率"。介电常数通常随温度和介质中传播的电磁波的频率而变。电容器用的电介质要求具有较大的介电常数，以便减小电容器的体积和重量。

电容式物位开关是利用无料时介质是空气，有料时介质是物料这种差别来实现料位报警的，空气的介电常数约为"1"，如果物料的介电常数越大，则无/有料之间的电容差别就越大，物位开关越好判断。而物料品种一旦固定了，它的介电常数也就固定了，要使无/有料的差别尽量大，只好延长探极长度，以使电容式物位开关稳定工作。当然，电容式物位开关也有"灵敏度标定"这种方法去适应上述"差别"的大、小，但灵敏度标定得太高，抗干扰能力会变低。所以，还是在可能情况下，用长一点探极好。

介电常数代表了电介质的极化程度，也就是对电荷的束缚能力，介电常数越大，对电荷的束缚能力越强。它是表示绝缘能力特性的一个系数，以字母 ε 表示，单位为法/米。

（1）绝对介电常数（absolute dielectric constant）ε_0，定义为 $1/\mu_0 C_0$，其中 μ_0 为真空磁导率，C_0 为光在真空中的速度；

（2）介电常数，定义为电通量密度 D 除以电场强度 E。其 SI 单位为法/米，常用微法/米、纳法/米、皮法/米；

（3）相对介电常数（relative dielectric constant）ε_r，定义为 $\varepsilon/\varepsilon_0$，其中 ε_0 为真空介电常数。它是无量纲量。在化工中一般使用相对介电常数来表征电介质或绝缘材料电性能。一般化工文献中，往往使用"介电常数"代替"相对介电常数"。但在有可能混淆的场合，不得把相对介电常数简称为介电常数。

Part 4　Translations

1. Apparently these sausages are electrostatically compatible with the iPhone's capacitive touch screen, making them an ideal finger alternative.

显然，这些香肠与 iPhone 电容触屏能"和谐共处"，还能作为最理想的手指替代品。

2. Families are no longer in close proximity to each other.

各家不再像以前一样比邻而居。

3. It contains a powerful and resonant truth: we want the most powerful among us to act with care and to wield that power responsibly.

它包含了一个强有力的、能引起共鸣的事实：我们希望最有权势的人谨慎行事，负责地行使其权力。

4. His voice sounded oddly resonant in the empty room.

他的声音在这空荡荡的房间里听起来异常嘹亮。

5. In modern physical science the opposition, first observed to exist in magnetism as polarity, has come to be regarded as a universal law pervading the whole of nature.

在近代自然科学里，最初在磁石里所发现的两极性的对立，逐渐被承认为浸透于整个自然界的普遍规律性。

Part 5 ▶▶ Reference Version

第七课 电容式接近传感器

电容式接近传感器的工作原理是基于在 RC 谐振电路中安装的电容器的电容量的变化来评价的（图 7-1）。当一个物体接近该传感器使得电容量增加。这导致了 RC 电路中的振荡变化，这一变化是可以测量的。电容的变化很大程度上取决于距离、尺寸和相应材料的介电常数。

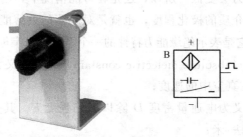

图 7-1 电容式接近传感器

电容式接近传感器能检测金属物体，也能检测非金属物体，对金属物体可以获得最大的动作距离，对非金属物体（如塑料、玻璃、水、油等物质）动作距离决定于材料的介电常数，材料的介电常数越大，可获得的动作距离越大。

接近传感器有一个 PNP 输出，即信号线被触发时切换到正电位。设计的开关为常开触点形式。负载连接在传感器的信号输出与地线之间。一种黄光发光二极管（LED）显示开关状态。电容式接近传感器不能平贴安装。

在操作过程中，请注意所施加的电压的极性。线缆终端彩色编码标示见表 7-1。

表 7-1 接线端颜色标识

参数	参数值
正极接线端	棕色
负极接线端	蓝色
负载的输出	黑色

电容式接近传感器的技术数据见表 7-2。

表 7-2 技术数据

参数	参数值
允许工作电压	10～55V 直流电
开关输出	PNP 或常开触点
正常开关距离	2～8mm
滞后（在正常开关距离）	3%～5%
最大开关电流	200mA
最大开关频率	300Hz
在 55V 状态下空载时的电流消耗	7mA
允许环境温度	20～70℃
保护等级	IP 65

Lesson 8 Ultrasonic sensor

Part 1 ▶▶ Text

In Fig. 8-1, the left Picture is the analog ultrasonic sensor layout, the light schematic diagram indicate the internal principle.

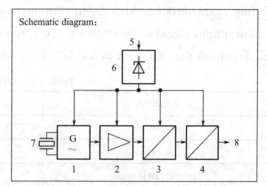

Fig. 8-1 Ultrasonic sensor

1 Oscillator

2 Amplifiers

3 Evaluating unit

4 Measuring transducer

5 External voltages

6 Internal constant power supplies

7 Ultrasonic converters with active zone

8 Output: Current signal

The operational principle of an ultrasonic sensor is based on the generation of acoustic waves and their detection following reflection on an object. Normally, atmospheric air acts as a carrier of the ultrasonic waves.

A sound generator is actuated for a short period of time and emits an ultrasonic pulse which is inaudible to the human ear. Following emission, the ultrasonic pulse is reflected on an object located within range and echoed back to the receiver. The duration of the ultrasonic pulse is evaluated electronically. Within a certain range, the output signal is proportional to the signal duration of the ultrasonic pulse.

The object to be detected can be made of different materials. The shape or colour, Function solid, fluid or powdery condition do not have any or a very minimal effect on detection. In the case of objects of smooth, even surface, the surface must be aligned vertically to the ultrasonic beam.

With this kind of sensor you are able to do two kinds of measurements: First you can measure the distance between the sensor and an object. The manufacturer setup of the sensor is ideal for this kind of measurement. Rising output signal at rising distance to the object.

But for measuring the filling level of a container a different setup is necessary because with a rising filling level the distance of the measured object (water surface) to the sensor is getting smaller. Therefore the signal output was changed from rising to falling characteristics. Also the measurement range was changed so that we can get maximum output signal at maximum and the minimum output signal at minimum filling level. Technical data isshown in Table 8-1.

Table 8-1 Technical data

Parameter	Value
Protection class	IP 67
Weight	max. 67g
Ambient temperature	−25 bis 70℃
Switching point error	±2.5%(−25 to 70℃)
Rated operation voltage U_e	24V DC
Operation voltage range U_B	20 to 30V DC(at 12 to 20V DC reduced sensitivity up to 20%)
Permissible residual ripple	10%
Idle current consumption I_0	<50mA
Switch output(NC/NO)/ Frequency output(FA) Rated operating current I_e Voltage drop U_d	≤150mA ≤3V at 150mA
Analog output(U_A/I_A) Current range Burden	4 to 20mA 0 to 300Ω
Sensor activ	Operating voltage or high impedance input current IE max. 16mA
Sensor not activ	0 to 3V Input current IE max 11mA
Measurement range	From:50mm To:345mm
Max. measurement range	From:46mm To:346mm
Output(current)	4 to 20mA

The oscillations at the beginning and the end of the characteristic curve (Fig. 8-2) are caused by the type of sensor construction. For the characteristic curve displayed, the distance between the sensor and the bottom of the container was adjusted to 330 mm.

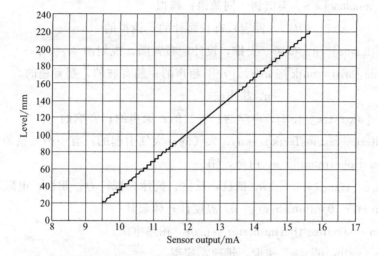

Fig. 8-2 Characteristic curve

Please notice that the sensor is not linear over the whole measurement range. You should use this sensor only for measurements in the tank with filling levels between 80mm to 180mm. outside this borders the sensor might differ from the characteristic curve.

Note

Please see the figure (Fig. 8-3) below to see the connectivity between output signal and distance.

- A distance >500mm is not defined for measurement.
- A value of 20mA is only theoretically measurable.

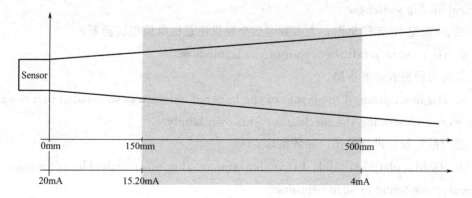

Fig. 8-3 Distance an measurement range

Part 2 New Words and Phrases

analog [ˈænəlɔːg] n. 类似物，同源语；模拟
　　　　　　　　　adj.（钟表）有长短针的；模拟的
layout [ˈleɪaʊt] n. 布局，安排，设计；布置图，规划图
ultrasonic [ˌʌltrəˈsɒnɪk] adj. [声] 超声的；超音波的，超音速的
　　　　　　　　　n. 超声波
acoustic [əˈkuːstɪk] adj. 声学的；听觉的；原声的；音响的
atmospheric [ˌætməsˈferɪk] adj. 大气的；大气引起的；有……气氛的
platinum [ˈplætɪnəm] n. 白金，铂
resistance [rɪˈzɪst(ə)ns] n. 抵抗，反抗；抗性；抵抗力；耐性；电阻
thermometer [θəˈmɒmɪtə(r)] n. 温度计，体温计
platinum resistance thermometer sensor　铂热电阻
velocity [vəˈlɒsəti] n. 速度，迅速，速率
thermo [ˈθɜːməʊ] adj. 热的，热电的
operational principle [ˌɒpəˈreɪʃənəlˈprɪnsəpl] 工作原理，操作原理

Part 3 Translations

1. One of the most atmospheric corners of Prague is the old Jewish ghetto.
古老的犹太人社区是布拉格最具特色的角落之一。

2. More evidence has emerged that the apparent slowdown in the rate of atmospheric warming may be explained by heat absorption in the deep ocean.
有更多证据表明，大气升温的速度明显放缓可能是因为深海吸收了热量。

3. In the early 1970s, the majority of telephone exchanges were still electromechanical analog switches.
在20世纪70年代早期，大多数电话交换机还是机电模拟转换器。

4. He tried to recall the layout of the farmhouse.
他努力回想农舍的布局。

5. His description of the layout of the Imperial Palace was so detailed that it was as if he were enumerating the heirlooms of his own family.
他对故宫设计讲解甚详，如数家珍。

6. Gold, platinum and diamonds, among the most valuable of metals and minerals, are found in such deposits.
最贵重金属和矿物中的金、铂和钻石就是在这种矿床中发现的。

7. Edison's version (pictured) was able to transcend others because of some key details, including finding an effective incandescent material and being able to achieve a higher vacuum and a high resistance.

爱迪生的发明版本能够从中胜出是由于一些关键的细节，包括找到了一种有效的发白热光的材料，灯泡内能够达到较高程度的真空和高电阻。

8. But volume is only the first dimension of the big data challenge; the other two are velocity and variety.

但是，数据量（volume）只是大数据挑战的一个方面，其他两个方面指的是速度（velocity）和多样性（variety）。

Part 4 ▶▶ Reference Version

第八课　超声波传感器

在图 8-1 中，左图是模拟量超声波传感器外观图，右图是内部原理示意图。

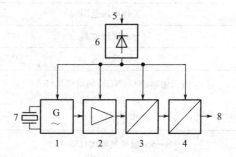

图 8-1　超声波传感器

1. 振荡器
2. 放大器
3. 检测元件
4. 测量转换元件
5. 外加电压
6. 内部持续电源
7. 有源区的超声波转换器
8. 输出的电流信号

超声波传感器的工作原理是基于声波的产生和检测物体的反射。通常情况下，大气是超声波的载体。声波发生器启动一小段时间，然后放出超声波脉冲，这是人耳听不到的。接下来，超声波脉冲反射到固定范围内的对象上并回荡到接收器。超声波脉冲的持续时间是电子评估的。在一定范围内，输出信号与超声波脉冲信号的持续时间是成比例的。

被检测的对象可以是由不同的材料制成的。形状或颜色，固体、液体或粉末状态对检测都没有影响或影响很小。如果遇到光滑表面的物体，表面必须垂直对准于超声波束。

用这种传感器我们可以做两种类型的测量。首先你可以测量传感器和目标物体之间的距离。制造商设置的该传感器对于这种测量是理想的。输出信号会随着与目标物体间的距离加大而上升。但是如果测量容器内的液位时，就需要一个不同的设置。因为随着液位的上升，被测物体到传感器的距离会变小。因此，输出信号就要随液位从上升到下降的特性而改变。同时测量范围也要改变，这样我们可以在最大液位获得最大的输出信号，在最低液位获得最小的输出信号。

表 8-1 技术数据

参量	参数值
保护等级	IP 67
重量	最大值 67g
环境温度	−25～70℃
开关点误差	±2.5%(−25～70℃)
额定工作电压	24V 直流电
工作电压范围	20～30V 直流(在 12～20V 直流时灵敏度下降到 20%)
允许残留波痕	10%
空载电流	<50mA
开关输出(NC /NO)/频率输出(FA) 额定工作电流 电压降落 U_D	≤150mA ≤3V at 150mA
Burden 模拟量输出(U_A/I_A) 电流范围 阻抗	4～20mA 0～300Ω
传感器激活	工作电压或高阻抗 输入电流最大 16mA
传感器未被激活	0～3V 输入电流最大 11mA
测量范围	50～345mm
最大测量范围	46～346mm
输出(电流)	4～20mA

该特性曲线（图 8-2）开始和结束部分的曲线是由传感器的结构引起的。从特性曲线上可以看出，传感器和容器底部之间的距离调整到了 330mm。请注意在整个测量范围内传感器不呈线性状态。因此这种传感器只能用于测量填充物在 80～180mm 的容器罐。在这个边界之外，传感器可能就与特性曲线的显示不同。

注意：

请参见图 8-3 所示输出信号和距离之间的连通性。

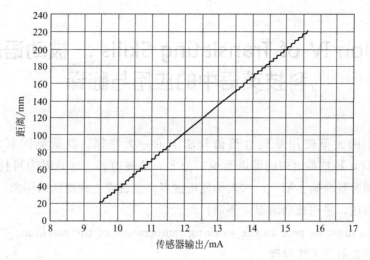

图 8-2 特性曲线

- 距离大于 500mm 的不在定义测量范围内。
- 20mA 仅在理论上是可测量的。

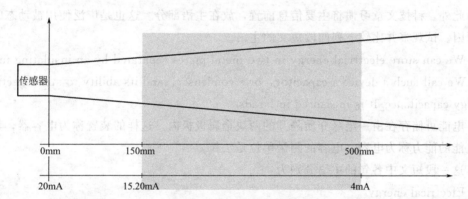

图 8-3 距离测量范围图

Section IV of Translating Skills：被动语态在科技英语中的应用与翻译

被动语态的大量使用是机电英语句法上的一大特征。根据英国利兹大学 John Swales 的统计，科技英语中的谓语至少三分之一是被动态。这是因为科技文章侧重叙事推理，强调客观准确。第一、二人称使用过多，会造成主观臆断的印象。因此尽量使用第三人称叙述，采用被动语态，例如：

Attention must be paid to the working temperature of the machine.

应当注意机器的工作温度。

而很少说：

You must pay attention to the working temperature of the machine.

你们必须注意机器的工作温度。

此外，科技文章习惯将主要信息前置，放在主语部分。这也是广泛使用被动态的主要原因。试观察并比较下列两段短文的主语。

We can store electrical energy in two metal plates separated by an insulating medium. We call such a device a capacitor, or a condenser, and its ability to store electrical energy capacitance. It is measured in farads.

电能可储存在由一绝缘介质隔开的两块金属极板内。这样的装置称为电容器，其储存电能的能力称为电容。电容的测量单位是法拉。

这一段短文中各句的主语分别为：

Electrical energy

Such a device

Its ability to store electrical energy

It（Capacitance）

它们都包含了较多的信息，并且处于句首的位置，非常醒目。四个主语完全不同，避免了单调重复，前后连贯，自然流畅。足见被动结构可收到简洁客观之效。

虽然英语和汉语都有被动语态，但两种文字对被动语语态的运用却不尽相同。同一个意思，英语习惯用被动语态表达，汉语却往往要用主动语态。英译汉时，必须注意被动语态的译法，不要过分拘泥于原文的被动结构，而要根据汉语的习惯，做灵活多样的处理。下面我们来学习几种被动语态的翻译方法。

一、译成汉语的主动句

1. 原主语仍译为主语（省略被字的被动句）

当英语被动句中的主语为无生命的名词，又不出现由介词 by 引导的行为主体时，往往可译成汉语的主动句，原句的主语在译文中仍为主语。

Sand casting is used to make large parts.

砂型铸造用于制造大型零件。

Materials may be grouped in several ways.

材料可以按几种方式分类。

2. 把原主语译成宾语,而把行为主体或相当于行为主体的介词宾语译成主语

Friction can be reduced and the life of the machine prolonged by lubrication.

润滑能减少摩擦,延长机器寿命。

A new kind of magnifying glass is being made in that factory.

那个工厂正在制造一种新的放大镜。

3. 增译逻辑主语

原句中未包含动作的发出者,译成主动句时可从逻辑主语出发,适当增添不确定主语,如"人们"、"大家"、"我们"等。

Salt is known to have a very strong corroding effect on metals.

大家知道,盐对金属有很强的腐蚀作用。

Magnesium is found to be the best material for making flash light.

人们发现,镁是制造闪光灯的最好材料。

二、译成汉语的无主句

汉语句的无主句与英语相比是一种独特的句型。英语的许多被动句不需要或无法讲出动作的发出者,因此往往可译成汉语的无主句,而把原句中的主语译成宾语。

Attention has been paid to the new measures to prevent corrosion.

已经注意到采取新的防腐措施。

In the watch making industry, the tradition of high precision engineering must be kept.

在钟表制造业中,必须保持高精度工艺的传统。

三、译成汉语的判断句

凡是着重描述事物的过程、性质和状态的英语被动句,与系表结构相近,往往可译成"是…的结构"。

The volume is not measured in square millimeters. It is measured in cubic millimeters.

体积不是以平方毫米计量的,它是以立方毫米计量的。

The first explosive in the world was made in China.

世界上最早的炸药是在中国制造的。

四、译成汉语的被动句

英语的某些着重被动动作的被动句,汉译时要译成被动句,以突出其被动意义。但汉语表达被动语态的方式很多,除了"被"字外,还时常使用"由"、"受"、"加以"、"为"、"挨"等。

This kind of steel is not corroded by air and water.

这种钢不会被空气和水腐蚀。

The instruction will be discussed briefly.

说明书将简单地加以讨论。

第四节 翻译小练习

认真分析句子，准确翻译下列句中的被动结构。

1. This cylindrical vessel is designed and made by ourselves.

2. Full use must be made of functions of the machine.

3. Laser technology may be taken as one of the most important advances achieved by scientists in recent years.

4. A synthetic material equal to that metal in strength has been created by our research center.

5. Electricity itself has been known to man for thousands of years.

6. The repair of the machine will be finished in a week.

7. Being very small, an electron cannot be seen by man.

8. A computer can be given information or orders in various ways.

9. If water is heated, the molecules move more quickly.

10. The degree of water pollution can be detected with the apparatus newly invented.

第四节 翻译小练习答案

1. 这种圆柱形容器是我们自行设计和制造的。
2. 必须充分利用机器的各项功能。
3. 激光技术可以看作是近年来科学家取得的最重要成就之一。
4. 我们的研究中心创造了一种强度与那种金属相同的合成材料。
5. 电为人类所知晓已有好几千年了。
6. 机器的修理将在一周后完成。
7. 因为电子很小，所以人们看不见。
8. 我们能以各种方式给计算机发送信息或指令。
9. 如果把水加热，分子运动就更快。
10. 采用这种新发明的仪器可以检测出水质污染的程度。

Lesson 9 • Temperature sensor

Part 1 ▶▶ Text

Temperature sensor (transducer temperature) is a sensor that can sense the temperature and convert it into usable output signal (Fig. 9-1). The temperature sensor is the core part of the temperature measuring instrument. According to the measurement method can be divided into two categories of contact and non-contact, according to the characteristics of the sensor materials and electronic components are divided into two types of thermal resistance and thermocouple.

Fig. 9-1 Temperature sensor

The detection part of the contact type temperature sensor has a good contact with the measured object, and the thermometer is also called a thermometer. A thermometer can achieve thermal balance by conduction or convection, so that the value of the thermometer can directly indicate the temperature of the object being measured. The sensor in the general measurement accuracy is higher. Within a certain range of temperature measurement, the thermometer can also measure the temperature distribution inside the object. But for the moving body, small targets or small heat capacity of the object will have a larger measurement error. These types of sensors are commonly used, for example, double metal thermometer, glass liquid thermometer, pressure type thermometer, resistance thermometers, thermistors and thermocouples. They are widely used in industry, agriculture, commerce and other departments. In everyday life people often use these thermometer.

Non contact sensitive components are not in contact with the measured object, also known as non contact type temperature measuring instrument. This instrument can be used to measure the surface temperature of moving objects, small targets, small heat capacity and temperature changes, and can also be used to measure the temperature distribution of the temperature field.

The most commonly used of non-contact temperature measuring instrument based on the basic law of the black body radiation, called radiation thermometer. Radiation thermometry include brightness method (see optical pyrometer) radiation (radiation pyrometer) and colorimetric method (see colorimetric thermometer). All kinds of radiation temperature measurement method can only measure the corresponding luminosity temperature, radiation temperature or specific color temperature. The temperature is measured only by the body of a black body (which absorbs all the radiation that does not reflect light). If you want to measure the true temperature of the object, it is necessary to modify the surface emissivity of the material. The surface emissivity of the material depends not only on the temperature and wavelength, but also on the surface state, coating and microstructure, so it is difficult to accurately measure. In the automatic production often requires use of radiation thermometry to measure or control the surface temperature of certain objects, such as the metallurgy steel rolling temperature, roll temperature, forging temperature and various molten metal in the smelting furnace or crucible temperature. In these circumstances, it is difficult to measure the surface emissivity of the object. For the automatic measurement and control of the temperature of the solid surface, the additional mirror can be used to form a black body cavity together with the measured surface. The effect of additional radiation can improve the effective radiation and effective emissivity of the tested surface. The effective emissivity coefficient is used to correct the measured temperature, and the real temperature of the measured surface can be obtained.

The temperature sensor (Fig. 9-1) contains a platinum resistance thermometer with interchangeable measuring element. The sensor consists of a shield tube, a connection head and the measuring element.

During installation, ensure as accurately as possible that the sensor accepts the temperature to be measured. Heat absorbed or given out by the sensor is to be avoided.

The temperature sensor is screwed into a threaded hole in the container.

Resistance default value of platinum resistance thermometer Pt100 as a function of temperature.

According to the measurement data shown in Table 9-1, draw the characteristic curve of the Pt100 sensor at minus 100 degrees Celsius to 200 degrees Celsius (Fig. 9-2).

Table 9-1 the measurement date of temperature sensor

Temperature/℃	−100.00	0.00	100.00	200.00
Basic value/Ω	60.25	100.00	138.50	175.84

The permissible flow velocity for water is 3m/s. To disassemble the sensor there is no need to remove the whole element out of the container. Just remove the two grub

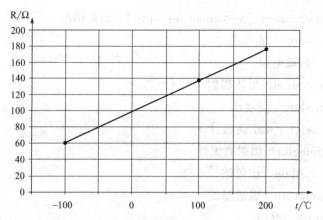

Fig. 9-2　Characteristic curve of the Pt100-sensor from $-100℃$ to $+200℃$

screws. After that you are able to remove the thermo element very easily.

Part 2　New Words and Phrases

platinum ['plætnəm]　n. 白金，铂
resistance [rɪ'zɪstəns]　n. 抵抗，反抗；抗性；抵抗力；耐性；电阻
thermometer [θə'mɒmɪtə(r)]　n. 温度计，体温计
velocity [və'lɒsəti]　n. 速度，迅速，速率
commerce ['kɒmɜːs]　n. 商务；商业；贸易；商业人员
luminosity [ˌluːmɪ'nɒsəti]　n. 光度；发光；光明；发光体
corresponding [ˌkɒrə'spɒndɪŋ]　adj. 相当的，对应的；符合的；一致的
　　　　　　　　　　　　　　v. 相符合（correspond 的现在分词）；类似（correspond 的 ing 形式）；相配
emissivity [ˌemɪ'sɪvɪti]　n. 发射率；放射性
metallurgy [mə'tælədʒi]　n. 冶金，冶金学，冶金术
molten metal　熔融金属；熔化金属；金属液；已熔金属
crucible ['kruːsɪbl]　n. 坩埚，严酷的考验
circumstance ['sɜːkəmstəns]　n. 环境，境遇；事实，细节；典礼，仪式

Part 3　Technical Words and Phrases

thermal resistance [ˈθəːməl riˈzistəns]　热阻，热变电阻，热敏电阻；
platinum resistance thermometer sensor　铂热电阻
thermocouple　热电偶
thermometer　温度计

measurement accuracy ［'meʒəmnent 'ækjurəsi］ 测量精度
measurement error 测量误差
thermistors 热敏电阻
black body radiation 黑体辐射
brightness method 亮度法
optical pyrometer 光学高温计
radiation pyrometer 辐射高温计
colorimetric method 比色法
forging temperature 锻造温度
roll temperature 辊温
smelting furnace 熔炉，冶炼炉，焙炼炉

Part 4 Translations

1. This much is certain: the tendency is upward with the march of civilization, toward a higher velocity.

有一点非常确定：随着文明的进步，流通速度将保持上升的趋势。

2. Wait for a couple of minutes with your mouth closed before inserting the thermometer.

先合上嘴巴等几分钟再放进体温计。

3. For small measurement sensor, it should be avoided under attack, this will affect its future measurement accuracy.

对于小计量的传感器，应该避免它受到冲击，这会影响到它今后测量值的准确性。

4. Experimental results show that this system can rival the system using thermistors for power detection in the measurement accuracy.

测量结果表明，该测试系统达到了与采用热敏电阻进行功率检测的测量系统相当的测量精度

5. The principle of over current protection of communication equipment is to make use of the positive temperature coefficient PTC thermistor to achieve the technical requirements.

通信设备过电流保护的原理是利用正温度系数 PTC 热敏电阻来达到技术要求。

6. Growth in electronic commerce is not a straight line trend&. it come slowly at first, then accelerates rapidly.

电子商务的发展并非一帆风顺，而是经历了缓慢起步才加速前进的。

7. The biggest impact has occurred where electronic commerce matches buyers and sellers who would not previously have found each other.

当电子商务把以前不可能找到对方的买方和卖方匹配起来时，网络的最大影响力才出现。

8. They have made their fortunes from industry and commerce.

他们靠工商业发了财。

9. In this situation astronomers tacitly assume that stars of the same luminosity and color all have the same mass.

在目前的状况下，天文学家心照不宣地默认，光度和颜色都相同的恒星具有相同的质量。

10. The experimental results show that the above-mentioned method can solve the true temperature and spectral emissivity measurement of most engineering materials.

实验结果表明，该方法可以解决绝大多数工程材料的目标真温及光谱发射率的测量问题。

11. In recent years, powder metallurgy is developing rapidly especially in the east China, promotion10% every year.

近年来，我国粉末冶金行业发展很快，东部和沿海地区的年增长幅度均在10%上。

12. You should soon accommodate yourself to the new circumstance.

你应尽快适应新环境。

Part 5 ▶▶ Reference Version

第九课　温度传感器

温度传感器（temperature transducer）是指能感受温度并转换成可用输出信号的传感器（见图9-1）。温度传感器是温度测量仪表的核心部分，品种繁多。按测量方式可分为接触式和非接触式两大类，按照传感器材料及电子元件特性分为热电阻和热电偶两类。

图9-1　温度传感器

接触式温度传感器的检测部分与被测对象有良好的接触，又称温度计。温度计通过传导或对流达到热平衡，从而使温度计的示值能直接表示被测对象的温度。该种传感器一般测量精度较高。在一定的测温范围内，温度计也可测量物体内部的温度分布。但对于运动体、小目标或热容量很小的对象，则会产生较大的测量误差，常用的温度计有双金属温度计、玻璃液体温度计、压力式温度计、电阻温度计、热敏电阻和温差电偶等，它们广泛应用于工业、农业、商业等部门。在日常生活中人们也常常使用这些温度计。

非接触式的敏感元件与被测对象互不接触，又称非接触式测温仪表。这种仪表可用来测量运动物体、小目标和热容量小或温度变化迅速（瞬变）对象的表面温度，也可用于测量温度场的温度分布。

最常用的非接触式测温仪表基于黑体辐射的基本定律,称为辐射测温仪表。辐射测温法包括亮度法(见光学高温计)、辐射法(见辐射高温计)和比色法(见比色温度计)。各类辐射测温方法只能测出对应的光度温度、辐射温度或比色温度。只有对黑体(吸收全部辐射并不反射光的物体)所测温度才是真实温度。如欲测定物体的真实温度,则必须进行材料表面发射率的修正,而材料表面发射率不仅取决于温度和波长,而且还与表面状态、涂膜和微观组织等有关,因此很难精确测量。在自动化生产中往往需要利用辐射测温法来测量或控制某些物体的表面温度,如冶金中的钢带轧制温度、轧辊温度、锻件温度和各种熔融金属在冶炼炉或坩埚中的温度。在这些具体情况下,物体表面发射率的测量是相当困难的。对于固体表面温度自动测量和控制,可以采用附加的反射镜使与被测表面一起组成黑体空腔。附加辐射的影响能提高被测表面的有效辐射和有效发射系数。利用有效发射系数通过仪表对实测温度进行相应的修正,最终可得到被测表面的真实温度。

本文图 9-1 中所示的温度传感器是包含一个带有可互换测量元件的铂电阻温度计。该传感器由屏蔽管、连接头和测量元件组成。在安装时要做到尽可能准确,确保传感器能接收被测温度。传感器吸收与放出的热量可忽略不计。温度传感器被拧入容器的螺纹孔中。铂电阻温度计 Pt100 电阻作为温度函数的默认值。

通过表 9-1 所示的测量数据,绘制出如图 9-2 所示的 Pt100 传感器在 -100~200℃ 的特性曲线。

表 9-1　温度传感器温度电阻测量值

温度/℃	-100.00	0.00	100.00	200.00
基本值/Ω	60.25	100.00	138.50	175.84

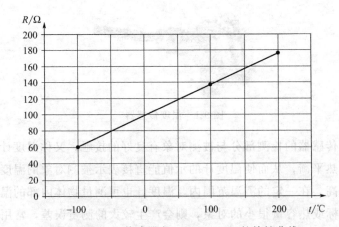

图 9-2　Pt100 传感器在 -100~200℃ 的特性曲线

水的允许流速为 3m/s。拆卸传感器时无需拆下容器的整个元件。只去掉两个平头螺钉即可。然后你就可以容易地删除热元件。

Section V of Translating Skills：定语从句在科技英语中的应用与翻译

在英语的各种从句中，定语从句最为复杂，因此翻译时难度也往往最大。我们知道，在句中作定语的从句，称为定语从句，由关系代词和关系副词引导，置于被修饰语之后。但现代汉语里，却没有关系代词和关系副词。英语中定语从句后置的这种语言现象，汉语里也是没有的。英语定语从句的后置使句子长而复杂。而汉语的定语一般是前置定语，句子较短。英汉两种语言句子结构上的这些差异给翻译定语从句带来了较大困难。有些定语从句不仅结构复杂，而且含义繁多，与主句之间存在着状语关系，具有补充、原因、转折、结果、条件等意义。在科技英语中，定语从句和被动语态一样，出现的频率很高。下面我们就根据定语从句的不同特征来阐述定语从句在科技英语中的翻译方法。

定语从句的译法（一）

一、合译法

所谓合译法，主要是指把英语定语从句译成汉语的"的"字结构，放在被修饰词前，从而把定语从句和主句合译成汉语的单句。合译法适合于翻译结构不很长的定语从句。例如：

1. Space and Oceans are the new world which scientists are trying to explore.

太空和海洋是科学家正在努力探索的新领域。

2. How to make the machine run normally is a pressing problem which we must deal with.

怎样使机器正常运行是我们必须解决的紧迫问题。

二、分译法

所谓分译法，是指将定语从句与主句分开，译成并列分句。凡形式上的定语从句，一般宜于译成并列分句。此外十分冗长的意义上的定语从句，译成汉语前置定语难于安排，或使句子冗长，头重脚轻，不符合汉语的表达习惯，也可采用分译法。例如：

1. Galileo was a famous Italian scientist by whom the Copernican theory was further proved correct.

伽利略是一个著名的意大利科学家，他进一步证明哥白尼学说是正确的。

2. They worked out a new method by which production has now been rapidly increased.

他们研制了一种新方法，这种新方法使生产大大提高。

三、融合法

融合法是指把原名中的主句和定语从句融合起来译成一个独立句子的译法。定语从句与先行词关系密切时，往往可以将主句与定语从句合并译成一个简单句，使定语从句在汉译后成为单句中的谓语等句子成分。此译法特别适用于"there be"结构带有定语从句的句型。例如：

1. There are some metals which possess the power to conduct electricity and the ability to be magnetized.

某些金属既能导电又能被磁化。

2. There have been good results in the experiment that have given him great encouragement.

实验中的良好结果给了他莫大的鼓舞。

定语从句的译法（二）

在英语中，有些从句从语法结构上看是定语从句，但跟主句在逻辑上却有状语关系，说明原因、结果、目的、时间、条件或让步等。因此，翻译时可按其逻辑关系及内含语义译成汉语各种相应的复合句，也就是汉语的原因、结果、目的、时间、条件或让步等复句。

一、带有状语意义的定语从句的译法

（一）译成表示原因的分句

例如：

1. You must grasp that concept of "work" which is very important in physics.

你必须掌握"功"的概念，因为它在物理学中很重要。

2. Einstein, who worked out the famous Theory of Relativity, won the Nobel Prize in 1921.

由于爱因斯坦提出了著名的"相对论"，因此，他于1921年获得了诺贝尔奖金。

（二）译成表示结果的分句

例如：

1. Copper, which is used so widely for carrying electricity, offers very little resistance.

铜的电阻很小，所以广泛地用来传输电力。

2. There was something original about the plan that pleased all of them.

这个方案富于创造性，因此他们都很喜欢．

（三）译成表示让步的分句

例如：

1. The scientist, who was dog-tired, went on with the experiment.

那位科学家，虽已精疲力竭，但还是继续进行实验。

2. There was something original about the plan that pleased all of them.

这个方案富于创造性，因此他们都很喜欢。

3. Thomas Edison, who had failed again and again in his experiment, did not lose heart.

尽管实验一再失败，托马斯·爱迪生并没有失去信心。

（四）译成表示条件的分句

例如：

1. For any machine whose input and output forces are known, its mechanical advantage can be calculated.

对于任何机器来说，如果知其输入力和输出力，就能求出其机械效益。

2. Those machine parts the surfaces of which are pitted, must be changed.

如果机器零件表面起了麻点，就必须更换。

二、特种定语从句的译法

所谓特种定语从句，指的是修饰整个主句内容的非限制性定语从句。这种定语从句只能由 which 和 as 引导。

（一）由 which 引导的特种定语从句

这种从句总是位于主句之后，通常说明整个主句，其前有逗号分开。英译汉时，一般采用分译法，which 通常译成"这"，有时也可译成"从而"、"因而"等。

例如：

1. Metal is very strong and can be made into any required shapes, which makes it possible to be widely used in industry.

金属坚硬，又能加工成任何所需形状，这使它能在工业中得到广泛应用。

2. Scientists are trying to imitate some of the capabilities of the human brain, which will be able to lead to the construction of an entirely new electronic computer.

科学家正试图模仿人脑的某些能力，这将能够导致建造一种完全新式的电子计算机。

（二）由 as 引导的特种定语从句

这种定语从句位置灵活，可以位于主句之后、之前，还可以位于主句当中，翻译时主句与从句分译，往往把 as 译成"正如……那样"、"如"、"像"等。常见的表达法及其译法：

  as seen from table 如表中所示
  as shown in the figure 如图所示
  as mentioned above 如上所述
  as is well known to all 众所周知
  as the name indicates 顾名思义

as may be imagined　可以想象

as has been pointed out　正如已指出的那样

as often happens　正如经常发生的那样

第五节 翻译小练习

说出下列句中的定语从句，并应用所学方法进行翻译。

1. Science plays an important role in the society in which we live.

2. In the room where the electronic computer is kept, there must be no dust at all.

3. A floating object displaces an amount of water whose weight equals that of the object.

4. There are some chemical fuels that are clean and smokeless.

5. Electronic computers, which have many advantages, cannot replace man.

6. Friction, which is often considered as a trouble, is sometimes a help in the operation of machines.

7. The company cannot afford to buy the equipment which is too expensive.

8. Matter has certain features that enable us to recognize.

9. Iron, which is not so strong as steel, finds wide application.

10. He wishes to clear the trouble as soon as possible that will attract public attention.

11. Energy can neither be created nor destroyed, which is the universally accepted law.

12. As is known to us all, the burning of any fuel requires oxygen.

第五节 翻译小练习答案

1. 物质科学在我们生活的社会中起着重要的作用。
2. 在存放电子计算机的房间里，不能有一点灰尘。
3. 浮动对象排除水的重量等于该对象的一个量。
4. 有些化学燃料是洁净而无烟的。
5. 电子计算机虽然有许多优点但它不能代替人。
6. 摩擦，常被认为是一种麻烦但有时它也有助于机器的运转。
7. 公司买不起这台设备，因为它太贵了。
8. 具有一定的特征能够使我们认识到它们。
9. 铁的强度不如钢，但它却有广泛的应用。
10. 他想尽快排除故障来吸引公众的注意。
11. 能量既不能创造也不能被消灭，这是公认的定理。
12. 众所周知，任何燃料的燃烧都需要氧气。

Unit 4

Pneumatics and Electropneumatics

Lesson 10 • Introduction Of Pneumatics and Electropneumatics

 Part 1 ▶▶ Text

Pneumatics is a section of technology that deals with the study and application of pressurized gas to produce mechanical motion. The first of the advantages of pneumatics is simplicity of design and control. Machines are easily designed using standard cylinders and other components, and operate via simple on-off control. The second of the advantages of pneumatics is reliability. Pneumatic systems generally have long operating lives and require little maintenance. Because gas is compressible, Equipment is less subject to shock damage. Gas absorbs excessive force, whereas fluid in hydraulics directly transfers force. Compressed gas can be stored, so machines still run for a while if electrical power is lost. The third of the advantages of pneumatics is Safety. There is a very low chance of fire compared to hydraulic oil. Machines are usually overloading safe.

Electropneumatics System are two different techniques of hybrid systems. Electropneumatics is successfully used in many areas of industrial automation. Production, assembly and packaging systems worldwide are driven by electropneumatic control systems. The change in requirements together with technical advances have had a considerable impact on the appearance of controls. In the signal control section, the relay has increasingly been replaced by the programmable logic controller in order to meet the growing demand for more flexibility. Modern electropneumatic controls also implement new concepts in the power section to meet the needs of modern industrial practice. Examples of this are the valve terminal, bus networking and proportional pneumatics.

Electropneumatic controllers have the following advantages over pneumatic control systems:

● Higher reliability (fewer moving parts subject to wear)

● Lower planning and commissioning effort, particularly for complex controls

● Lower installation effort, particularly when modern components such as valve terminals are used

● Simpler exchange of information between several controllers

Electropneumatic controllers have asserted themselves in modern industrial practice and the application of purely pneumatic control systems is a limited to a few special applications.

Pneumatics deals the use of compressed air. Most commonly, compressed air is used to do mechanical work-that is to produce motion and to generate forces. Pneumatic drives have the task of converting the energy stored in compressed air into motion.

Cylinders are most commonly used for pneumatic drives, for example, a linear cylinder and a rotary motor showing in Fig. 10-1. They are characterized by robust construction, a large range of types, simple installation and favorable price and performance. As a result of these benefits, pneumatics is used in a wide range of applications.

Fig. 10-1 Pneumatic linear cylinder and pneumatic swivel cylinder.

Some of the many applications of pneumatics are handling of workpieces (such as clamping, positioning, separating, stacking, rotating)

● Packaging
● Filling
● Opening and closing of doors (such as buses and trains)
● Metal-forming (embossing and pressing)
● Stamping

Part 2 ▶▶ New Words and Phrases

hybrid ['haɪbrɪd]　　*n.* 杂种；杂种动物；杂交植物；合成物；混合动力；混合式；混合型

　　　　　　　　　　adj. 杂种的；混合的；混合语的

Unit 4 Pneumatics and Electropneumatics

pneumatic [njuːˈmætɪk] *adj.* 空气的，气动的，有气胎的；此文中意为气动，是"气压传动与控制"（又称"气动技术"）的简称

preface [ˈprefəs] *n.* 序文，前言，绪言
v. 为……加序言；作为……的开端

production [prəˈdʌkʃ(ə)n] *n.* 生产；摄制；制作；演出；产量；产品；生产部

assembly [əˈsemblɪ] *n.* 装配；议会；集合；组合体；组装；组件；程序集

packaging [ˈpækɪdʒɪŋ] *n.* 包装

requirement [rɪˈkwaɪəm(ə)nt] *n.* 需求，要求，必要条件

considerable [kənˈsɪd(ə)rəb(ə)l] *adj.* 相当的，重要的，可观的

impact [ɪmˈpækt] *n.* 冲击，碰撞，撞击；影响；冲击力，撞击力；作用
v. 挤入；压紧；撞击；冲击，碰撞，撞击；产生影响

appearance [əˈpɪər(ə)ns] *n.* 外表，出现，登台

signal [ˈsɪɡn(ə)l] *n.* 信号；信号器；暗号；交通指示灯
v. 向……作信号，用信号通知，标志；发信号，打信号

demand [dɪˈmænd] *n.* 要求；需要；需求
v. 要求，请求；查问，盘诘；需要；传唤

flexibility [ˌfleksəˈbɪləti] *n.* 易曲性；弹性；适应性，灵活性

implement [ˈɪmplɪm(ə)nt] *n.* 工具；器具
v. 实现，执行，使生效

concept [ˈkɒnsept] *n.* 观念；概念

reliability [rɪlaɪəˈbɪləti] *n.* 可靠；可靠程度；可信赖性

commission [kəˈmɪʃ(ə)n] *n.* 佣金；任务，职权，权限；委任，委托；委员会
v. 委任，使服役，委托制作

compressed [kəmˈprest] *adj.* 被压缩的；扁平的

mechanical [mɪˈkænɪk(ə)l] *adj.* 机械的，力学的，机械性的

motion [ˈməʊʃ(ə)n] *n.* 运动；动作
v. 向……打手势；向……摇头示意；打手势；摆动，走；摇头示意

cylinder [ˈsɪlɪndə(r)] *n.* 圆筒，圆筒状之物；硬盘里有形存储单位（计算机用语）；气缸

linear [ˈlɪnɪə] *adj.* 线的，线状的，直线的

swivel [ˈswɪv(ə)l] *n.* 转环，旋转椅的台座，旋转轴承
v. 使旋转；给……装上转体；使转动；旋转；转动

robust [rəʊˈbʌst] *adj.* 强健的；健全的；茁壮的；结实的，坚固耐用的

characterize [ˈkerəktəraɪz] *v.* 表示……的特色，赋予特色

construction [kənˈstrʌkʃ(ə)n] *n.* 建筑；解释；建筑物；结构，构造

installation [ˌɪnstəˈleɪʃ(ə)n] *n.* 安装；就职；装置

favorable [ˈfeɪv(ə)rəb(ə)l]　*adj*. 赞成的；赞许的；有利的
performance [pəˈfɔːm(ə)ns]　*n*. 履行，成绩，执行
benefit [ˈbenɪfɪt]　*n*. 利益，好处；津贴，救济金；优势；义演，义卖
　　　　　　　　　v. 有益于；有助于；得益，受惠
workpiece [ˈwɜːkˌpiːs]　*n*. 件；工件；工件壁厚；加工件

Part 3　Technical Words and Phrases

industrial automation　工业自动化；工业自动化机器；工业自动化模块
electropneumatic　电动气压的；电动气动式的；电气气动
technical advance　技术进步
impact on　影响；对……冲击，碰撞；对……之影响
programmable logic controller　可编程逻辑控制器
power section　动力部分，电源部分
valve terminal　阀岛（是将多个阀及相应的气控信号接口、电控信号接口甚至电子逻辑器件等集成在一起的一种集合体）
bus networking　总线
proportional pneumatic　比例气动
compressed air　压缩空气
pneumatic drives　气动驱动装置
as a result of　作为结果
be used to　过去习惯于
such as　如此……的，使……那样的；例如

Part 4　Translations

1. Application of Computer-control Automatic Electropneumatic Brake in China is widely used.
微控自动式电气制动在我国的应用非常广泛。

2. Transfer arms have become symbols of the industrial automation.
机械手已经成为了工业自动化的标志。

3. The technological process is one straight line production, quick, convenient, and easy to control.
工艺流程为一条线式，生产速度快，可靠性高，方便操作和管理。

4. There was one especially absent-minded young man in the assembly line who sewed on buttons.

在装配线上有个缝纽扣的年轻人特别心不在焉。

5. In order to keep technical advance, it increased the science and technology devotion, owned a young and professional technology team.

为保持技术领先,公司不断加大科研投入,拥有一支年轻化、专业化的技术队伍。

6. The duty of this course is to study of Position Control Precision with Pneumatic Proportional Valve.

这节课的任务是研究气动比例阀位置控制精度。

Part 5 ▶▶ Reference Version

第十课　气动和电气气动技术简介

气动技术是以压缩机为动力源,以压缩空气为工作介质,进行能量传递信号传递的工程技术。气动技术的优点之一是设计和控制简单。使用标准气缸和其他气动元件使得设计机器设备较为简单,通过开关控制易于操作。气动技术的优点之二是可靠。气动系统一般都具有较长的使用寿命且只需很少量的维护。由于气体具有可压缩性,设备较少有冲击损伤问题。气体能够吸收过载力,而液压系统却只能直接传送作用力。压缩气体可被存储,因此,当断电时,机器仍然能够运行。气动技术的优点之三是安全。相比于液压油而言,起火的概率要小得多。机器通常能够做得过载保护。

电气气动系统是两种不同的混合动力系统技术。电气气动技术已成功地应用于许多工业自动化领域,对世界范围的自动化生产领域的生产、装配和包装系统的电气控制系统驱动产生了极大的影响。在信号控制部分,继电器已被越来越多的可编程逻辑控制器所取代,以满足日益增长的需求并提供更多的灵活性。现代电气气动控制技术在电源集成部分提出了一种新概念,后者更能契合和满足现代工业的实际需要。这方面的例子是阀终端总线网络和比例气动。

电气气动控制技术较传统气动控制系统具备如下优点：
- 可靠性高（较少的运动部件磨损）
- 更少的规划和调试工作,尤其是复杂的控制
- 更小的安装成本,特别使用现代部件如阀终端
- 控制器间信息的简单交换

电气气动控制技术的普及在现代工业实践中得以证明,纯气动控制系统局限在了有限的特殊应用场合中。

气动元件利用压缩空气来进行工作。最常见的是,压缩空气被用来做机械工作,即产生运动和能量。气动驱动装置的任务就是将压缩空气中储存的能量转化为运动。

气缸是最常用的气动驱动器,例如图10-1所示的直线型气缸和旋转型气马达。其特点是结构稳定、类型多、安装简单、价格低廉、性能优异。由于气缸的这些优点,使得气动技术获得了广泛的应用。

图 10-1 直线型气缸和旋转型气马达

气动技术常常应用在批量程序处理工件的场合（如夹紧、定位、分离、叠加、旋转）。例如：

- 包装
- 填充
- 打开和关闭的门（如公共汽车和火车）
- 金属成形（压花、压）
- 冲压

Lesson 11 Basic control engineering terms

Part 1 ▶▶ Text

Pneumatic drives can only do work usefully if their motions are precisely carried out at the right time and in the right sequence. Coordinating the sequence of motion is the task of the controller. Control engineering deals with the design and structure of controllers.

The following section covers the basic terms used in control engineering.

Controlling-open loop control-is that process taking place in a system whereby one or more variables in the form of input variables exert influence on other variables in the form of output variables by reason of the laws which characterize the system. The distinguishing feature of open loop controlling is the open sequence of action via the individual transfer elements or the control chain. The term open loop control is widely used not only for the process of controlling but also for the plant as a whole.

For example, there is a device which closes metal cans with a lid (see Fig. 11-1). The closing process is triggered by operation of a pushbutton at the workplace. When the pushbutton is released, the piston retracts to the retracted end position. In this control, the position of the pushbutton (pushed, not pushed) is the input variable. The position of the pressing cylinder is the output variable. The loop is open because the output variable (position of the cylinder) has no influence on the input variable (position of the pushbutton). Controls must evaluate and process information (for example,

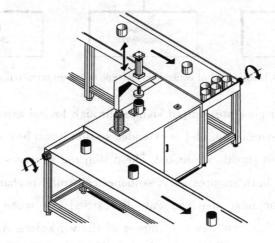

Fig. 11-1 Assembly device for mounting lids on cans

pushbutton pressed or not pressed). The information is represented by signals.

A signal is a physical variable, for example:
- The pressure at a particular point in a pneumatic system
- The voltage at a particular point in an electrical circuit

A signal is the representation of information. The representation is by means of the value or value pattern of the physical variable. An analog signal is a signal in which information is assigned point by point to a continuous value range of the signal parameter. In the case of a pressure gauge, each pressure value (information parameter) is assigned a particular display value (=information). If the signal rises or falls, the information changes continuously. A digital signal is a signal with a finite number of value ranges of the information parameter. Each value range is assigned a specific item of information. A pressure measuring system with a digital display shows the pressure in increments of 1 bar. There are 8 possible display values (0 to 7 bar) for a pressure range of 7 bar. That is, there eight possible value ranges for the information parameter. If the signal rises or falls, the information changes in increments.

A binary signal is a digital signal with only two value ranges for the information parameter. These are normally designated 0 and 1. A control lamp indicates whether a pneumatic system is being correctly supplied with compressed air. If the supply pressure (= signal) is below 5 bar, the control lamp is off (0 status). If the pressure is above 5 bar, the control lamp is on (1 status). Controllers can be divided into different categories according to the type of information representation, into analogue, digital and binary controllers (see Fig. 11-2).

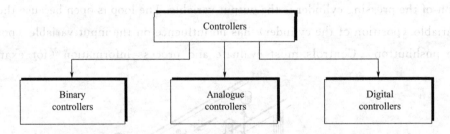

Fig. 11-2 Classification of controllers by type of information representation

A logic controller generates output signals through logical association of input signals. The assembly device in fig. 11-1 is extended so that it can be operated from two positions. The two output signals are linked. The piston rod advances if either pushbutton 1 or 2 is pressed or if both are pressed. A sequence controller is characterized by its step by step operation. The next step can only be carried out when certain criteria are met. Drilling station. The first step is clamping of the workpiece. As soon as the piston rod of the clamping cylinder has reached the forward end position, this step has been

Unit 4
Pneumatics and Electropneumatics

completed. The second step is to advance the drill. When this motion has been completed (piston rod of drill feed cylinder in forward end position), the third step is carried out, etc. A controller can be divided into the functions signal input, signal processing, signal output and command execution. The mutual influence of these functions is shown by the signal flow diagram (see Fig. 11-3).

- Signals from the signal input are logically associated (signal processing). Signals for signal input and signal process are low power signals. Both functions are part of the signal control section.

- At the signal output stage, signals are amplified from low power to high power. Signal output forms the link between the signal control section and the power section.

- Command execution takes place at a high power level-that is, in order to reach a high speed (such as for fast ejection of a workpiece from a machine) or to exert a high force (such as for a press). Command execution belongs to the power section of a control system.

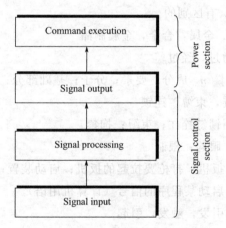

Fig. 11-3 Signal flow in a control system

Part 2 ▶▶ New Words and Phrases

usefully ['juːsfəli] adv. 有用地；有效地

precisely [prɪ'saɪsli] adv. 精确地，准确地；严格地，一丝不苟地；清晰地，明确地；刻板地，过分拘泥细节地

motion ['məʊʃ(ə)n] n. 运动；动作
　　　　　　　　　　v. 向……打手势；向……摇头示意；打手势；摆动，走；摇头示意

sequence ['siːkwəns] n. 序列，顺序，继起的事

coordinate [kəʊˈɔːdɪneɪt]　n. 同等者；坐标；同等物
　　　　　　　　　　　　v. 调整；整理；协调一致，协同动作；归于同一类别；成为同等重要
　　　　　　　　　　　　adj. 同等的；并列的

structure [ˈstrʌktʃə]　n. 结构，建筑物，构造
　　　　　　　　　　　v. 建筑，组织，构成

whereby [weə(r)ˈbaɪ]　adv. 凭什么，为何，借以，凭此，由于

variable [ˈveərɪəb(ə)l]　n. 变数，可变物，变量
　　　　　　　　　　　　adj. 可变的，易变的，不定的

exert [ɪgˈzɜːt; eg-]　v. 发挥，施以影响，运用

influence [ˈɪnfluəns]　n. 影响；势力；感化
　　　　　　　　　　　v. 影响；改变

characterize [ˈkerəktəraɪz]　v. 表示……的特色，赋予特色

distinguish [dɪˈstɪŋgwɪʃ]　v. 区别；辨认出；识别；把……区别分类；区别；辨别；识别

distinguishing　adj. 有区别的

metal [ˈmet(ə)l]　n. 金属，合金，金属制品

via　prep. 经由，通过，经过

element [ˈelɪm(ə)nt]　n. 成分；要素；分子；基础部分；典型部分

chain [tʃeɪn]　n. 链，束缚，连锁
　　　　　　　　v. 用锁链拴住；束缚；拘禁

lid [lɪd]　n. 盖子；眼睑；限制

trigger [ˈtrɪgə]　n. 扳机，打枪要拉起的扳机；启动装置；触发器；扳柄；计算机启动某程序的信号（计算机用语）
　　　　　　　　v. 引发，触发，引起

piston [ˈpɪst(ə)n]　n. 活塞，瓣

retract [rɪˈtrækt]　v. 缩回，缩进；撤回，收回

mounting [ˈmaʊntɪŋ]　n. 上马；可骑的东西；装备；升起；基础，架设；骑马；置放；安装，镶嵌一个新硬件（计算机用语）

represent [ˌreprɪˈzent]　v. 描绘，表现；表示；象征；作为……的代表

physical [ˈfɪzɪk(ə)l]　n. 体格检查
　　　　　　　　　　　adj. 身体的，自然的，物质的

particular [pəˈtɪkjələ(r)]　n. 个别项目，详细说明
　　　　　　　　　　　　　adj. 特别的，挑剔的，独有的

voltage [ˈvəʊltɪdʒ]　n. 电压，伏特数

finite [ˈfaɪnaɪt]　adj. 有限的；[语] 限定的；[数] 有穷的，有限的
　　　　　　　　　n. 有限性；有限的事物

Part 3 ▶▶ Technical Words and Phrases

carry out　完成；贯彻；实现
Control engineering　控制工程；控制工程学
deal with　与……交涉，交易；处理；应付；关于
open loop control　开环控制
take place　发生；举行
in the form of　以……的形式
end position　终端位置
electrical circuit　电路
pneumatic system　气动系统

Part 4 ▶▶ Translations

1. Once we carry out the measurements, they will surprised how much information can be obtained.

一旦我们完成了测量工作，他们将会对能获得如此多的信息感到惊讶。

2. Therefore, analysis and design of NCS has been a hot research problem in the control theory and control engineering area.

因此，网络控制系统（NCS）的分析与设计成为目前控制理论与控制工程领域的热点研究课题。

3. Indeed, practically every carmaker was showing off some sort of hybrid motor.

事实上，每家厂商都在炫耀他们的一些混合动力车。

4. He made a deal with Harry, and a deal's a deal.

他和哈里作了一笔交易，交易是不能随便改的。

5. A baby can't easily coordinate his movement.

婴儿不易协调自己的动作。

6. Maybe we can coordinate the relation of them.

或许我们可以调和他们之间的关系。

7. Stress may act as a trigger for these illnesses.

压力可能会成为引发这些疾病的原因。

8. Unresolved or unacknowledged fears can trigger sleepwalking.

残留在心里的或不为人知的恐惧都可能引发梦游。

9. The market situation is difficult to evaluate.

市场状况难以评价。

10. We didn't originally plan for this, so we need to evaluate what type of adjustment is needed.

我们最初没有对此进行计划，因此我们需要评估进行什么类型的调整。

11. The physical universe is finite in space and time.

物质世界在时间和空间上是有限的。

12. The audience of this concert is a finite body.

这场音乐会的听众人数有限。

13. There are a finite number of possibilities, so this step is not as complicated as you might think.

由于可能性有限，这一步骤不会像您想象的那么复杂。

Part 5 ▶▶ Reference Version

第十一课　工程项目基本控制

　　气动元件要实现正确有用的动作，必须在正确的时间里，按照正确的步骤使其触发。协调触发气动元件运动的动作序列是控制器的任务。控制工程来实现控制器结构和原理的设计。

　　下文的介绍涵盖了控制工程经常使用的基本术语。

　　控制——开环控制。控制系统的开环控制过程中，存在一个或多个输入变量，依照某种控制规律对输出变量产生影响，而这种控制规律表征了系统的特征。开环控制的显著特征是通过特定传递元件或控制链，实现了开环动作。开环控制这一术语不仅广泛应用在某一控制过程中，而且常常作为一个整体术语而使用。

　　例如，这里有一个用来给金属罐压上瓶盖的装置（见图11-1）。压盖过程由操作一个设置在工位上的按钮来触发。当按钮被释放时，活塞缩回在缩回端位置。在这个控制过程中，该按钮的位置（被压下，还是被释放）是输入变量。气缸活塞杆的

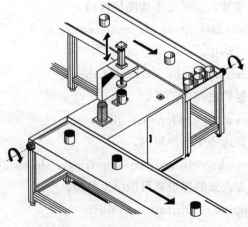

图11-1　金属罐压盖装置

位置是输出变量。这个控制回路是开环的原因是：因为输出变量（气缸的位置）对输入变量没有影响（按钮的位置），被控元件必须判断和处理输入信息（例如，按钮按下或未按下），这些信息来自于各种信号。

信号是一个物理变量，例如：
- 气动系统中的某一点压力
- 电路中某一点的电压

信号是用来表示某种信息参数的形式，是指物理变量的某个数值或数值的某种物理模式。某个模拟信号值代表的信息对应了某个物理量的特定值，这个物理量的特点是其参数具备一个连续的范围。例如，压力表的每个压力值（信息参数）被指定一个特定的显示值（＝信息）。如果信号上升或下降，那么压力表代表的信息参数即压力值不断变化。数字信号是一种具有有限数量的信息参数的值范围的信号。每个值的范围代表了信息的特定项目。带有数字显示的压力测量系统显示了 1 巴增量的压力。有 8 个可能的显示值（0～7 巴），则压力范围为 7 巴。也就是说，有 8 个可能的值，这个范围即为信息参数。如果信号上升或下降，信息的增量随之变化。

二进制信号是一个数字信号，其代表的信息参数只有两个值。这些值通常是指定的 0 和 1。例如，某控制指示灯用来指示一个气动系统是否正确供给压缩空气。如果供应压力（＝信号）低于 5 巴，控制灯关闭（0 状态），如果压力超过 5 巴，控制灯打开（1 状态）。控制器可以根据信息表示的类型分为不同的类型，分为模拟、数字和二进制控制器（见图 11-2）。

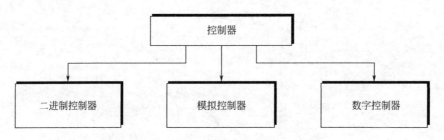

图 11-2　根据信息类型划分的不同控制器

逻辑控制器通过输入信号的逻辑关联来生成输出信号。图 11-1 中的装配装置可以被扩展，使其可以从 2 个不同的工作位置上被触发操作。2 个输出信号被相关联。如果按钮 1 或按钮 2 被压下或二者都被压下，那么活塞杆被推出。一个序列控制器的特点是它需要被一步一步操作，下一步只能在某些条件满足时进行。例如，一个钻井站。第一步是工件夹紧。当活塞杆的夹紧缸已经达到了前向的位置，表明这一步已经完成。第二步是推进钻头。当这项运动已经完成（进料缸活塞杆端位置），第三步才开始进行，等等。一个控制器的工作过程可以分为几个阶段，即功能信号输入、信号处理、信号输出和命令执行阶段。可以用信号流程图来表示这些功能的相互影响（见图 11-3）。

- 信号输入扫描阶段的信号来自输入信号的逻辑关联（输入后即进行信号处理）。信号输入和信号处理两个阶段的信号均是低功率信号。这两种功能都是信号控制部分的

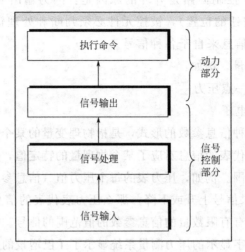

图 11-3 控制器工作过程流程图

一部分。

●在信号输出阶段，信号从低功率被放大到大功率。信号输出构成信号控制部分和电源部分之间的连接。

●命令的执行需要在高功率回路部分完成，例如，为了达到一个较高的速度（如从机工件快速喷射）或产生较高的力（如压力）。命令执行属于控制系统的电源回路部分。

Lesson 12 Pneumatic and electropneumatic control systems

Part 1 ▶▶ Text

The components in the circuit diagram of a purely pneumatic controller are arranged so that the signal flow is clear. Bottom up: input elements (such as manually operated valves), logical association elements (such as two-pressure valves), signal output elements (power valves, such as 5/2-way valves) and finally command execution (such as cylinders).

Both pneumatic and electropneumatic controllers have a pneumatic power section (see Fig. 12-1 and Fig. 12-2). The signal control section varies according to type.

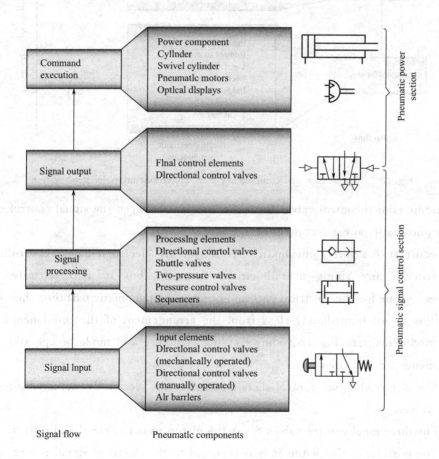

Fig. 12-1 Signal flow and components of a pneumatic control system

- In a pneumatic control pneumatic components are used, that is, various types of valves, sequencers, air barriers, etc.
- In an electro-pneumatic control the signal control section is made up of a electrical components, for example with electrical input buttons, proximity switches, relays, or a programmable logic controller.

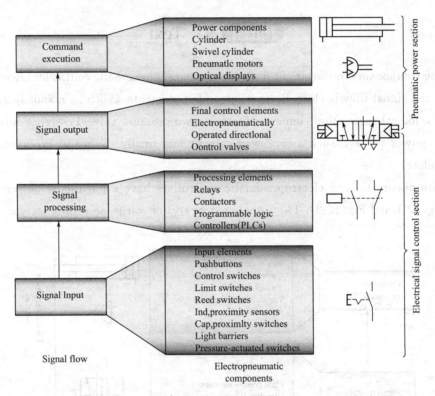

Fig. 12-2　Signal flow and components of an electropneumatic control system

The directional control valves form the interface between the signal control section and the pneumatic power section in both types of controller.

In contrast to a purely pneumatic control system, electropneumatic controllers are not shown in any single overall circuit diagram, but in two separate circuit diagrams——one for the electrical part and one for the pneumatic part. For this reason, signal flow is not immediately clear from the arrangement of the components in the overall circuit diagram. Fig. 12-3 shows at the structure and mode of operation of an electropneumatic controller.

- The electrical signal control section switches the electrically actuated directional control valves.
- The directional control valves cause the piston rods to extend and retract.
- The position of the piston rods is reported to the electrical signal control section by proximity switches.

Unit 4
Pneumatics and Electropneumatics

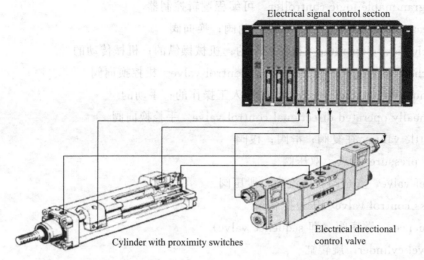

Fig 12-3　Structure of a modern electropneumatic controller

Part 2　▶▶ New Words and Phrases

vary [ˈveərɪ]　v. 改变，使多样化，变更；变化，违反，不同
proximity [prakˈsɪmətɪ]　n. 接近，亲近
relay [ˈriːleɪ]　n. 接力赛；接班的人（或动物）；轮换者；中继设备；继电器
　　　　　　　　v. 接转；播放；转发；转播
execution [ˌeksɪˈkjuːʃ(ə)n]　n. 实行；履行；执行；完成；执行机构
actuate [ˈæktʃueɪt]　v. 开动（机器、装置等）；激励；驱使；驱动；作用；促动
electrically [ɪˈlektrɪkəlɪ]　adv. 电力地；有关电地
extend [ɪkˈstend]　v. 延长，延伸；伸，伸出；扩大，扩展；致；伸展；延续；扩大
retract [rɪˈtrækt]　v. 缩回，缩进；撤回，收回；缩回，缩进；撤销，收回
according to　根据；取决于；按照；据……所载
that is　那就是
made up　补足；拼凑
made up his mind　下决心
in contrast with　与……相比

Part 3　▶▶ Technical Words and Phrases

pneumatic components　气动元件
proximity switch　接近开关

programmable logic controller 可编程逻辑控制器
directional control valve 方向控制阀；换向阀
mechanically operated 机械控制的；机械操纵的；机械传动的
mechanically operated directional control valve 机控换向阀
manually operated 手动操作的；人工操作的；手动的
manually operated directional control valve 手控换向阀
shuttle valve 往复阀；滑阀；梭阀
two pressure valve 双压阀
relief valve 安全阀；溢流阀；泄压阀
press control valve 压力控制阀
sequencer 顺序阀（即 sequence valve）
swivel cylinder 旋转缸
pneumatic motor 气动电动机；气马达
optical display 光电显示器
pressure actuated switch 压力驱动开关
light barriers 光栅
Cap. proximity sensor（capacitive proximity sensors） 电容接近传感器
Ind. proximity sensors（Inductive proximity sensors） 感应式接近传感器
reed switch 簧片开关；磁簧开关；干簧管；舌簧开关
limit switch 行程开关；限位开关；限动开关；微动开关
relay 继电器
contactor ['kɒntæktə] 接触器
electropneumatically operated directional control valve 电磁换向阀
piston rod 活塞杆

Part 4 ▶▶ Translation

1. In case air leakage occurs, please repair or replace the pneumatic components which may cause the leakage without any delay.

如有漏气现象，应及时修理或更换有关气动元件。

2. "The man who has made up his mind to win," said Napoleon, "Will never say: Impossible."

"一个决心赢的人，"拿破仑说，"从不会说不可能的。"

3. Proprietary features of a programmable controller that allow it to perform logic not normally found in relay ladder logic.

可编程控制器上独有的功能部件，使之可执行在继电器梯形逻辑中一般没有的逻辑功能。

Unit 4 Pneumatics and Electropneumatics

4. NC machine tools work so fast that their production rate is equal to the rate of five manually operated machine tools.

数控机床工作迅速,其生产率是手动机床的五倍。

5. Drag onto the page. Right-click to make a priority shuttle valve.

拖到绘图页上。右击可成为顺序往复阀。

6. No relief valve is required where two pressure reducing valves are installed.

如果装上两个减压阀,就不需要安全阀。

7. Adjustable relief valve, may be caused by non normal operation of hydraulic system.

任意调整溢流阀后,可能会造成液压系统非正常运转。

8. Pressure control valve is divided into benefits flow valve (safety valve), pressure relief valve, sequence valve, pressure relays, etc.

压力控制阀又分为溢流阀(安全阀)、减压阀、顺序阀、压力继电器等。

9. TURCK is a leading manufacturer of Inductive Proximity Sensors, Capacitive Proximity Sensors, Connectors, Cables, Cordsets, Field buses and Automation Controls.

图尔克在全世界的感应式接近传感器、电容式接近传感器、连接器、电缆、电线组件、现场总线和自动化控制产品制造商中处于领先地位。

10. Remark: When selecting braking unit, main contactor shall be chosen also. 备注:选择制动单元时必须同时配上主接触器进行保护。

11. Thermal relay value should be able to in no more than 140% of the rated current compressor when disconnect.

热继电器的设定值应能在不超过压缩机额定电流的140%时断开。

12. Most protection switches provide Auto, Manual or Remote relay switching between two or more channels.

大多数保护开关为自动、手动或多通道远程继电器开关。

Part 5 ▶▶ Reference Version

第十二课 气动和电气气动控制系统

在一个纯气动控制回路中,各个组成元件按照信号流的方向排列,结构很明确。自下而上分别为:输入部分(如手动控制阀)、逻辑元件(如双压阀)、信号输出单元(功率阀,如5/2阀)和最后执行元件(如气缸)。

无论是纯气动控制回路,还是电气气动控制回路,都有一个气动部分(见图12-1、图12-2)。信号控制部分根据类型不同而不同。

● 气动控制部分需要使用气动组件,即各类阀门、顺序阀、空气过滤组件,等等。

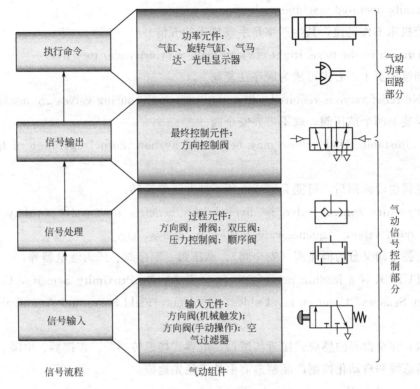

图 12-1　气动系统信号流程和元件

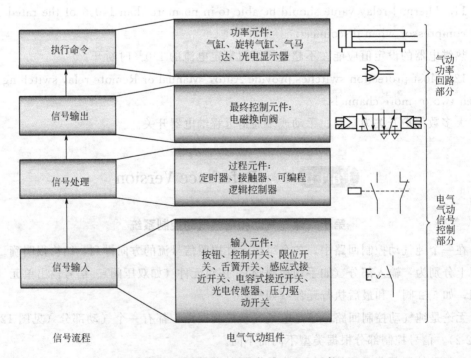

图 12-2　电气气动系统信号流程和元件

- 在电-气动控制信号的控制部分是由电气元件组成，例如电气输入按钮、接近开关、继电器、或可编程逻辑控制器等等。

方向控制阀在这两个控制回路中作为信号控制部分和气动功率部分间的接口。

与一个纯气动控制系统相反，一个电气气动控制回路无法在任何单一的整体回路图中表示，需在两个单独的回路图：一个电气部分和一个气动部分。基于这个原因，信号流程是不能清晰地从整体回路图元件布局表述出来的。图12-3显示了一个电气气动控制系统的结构和操作模式。

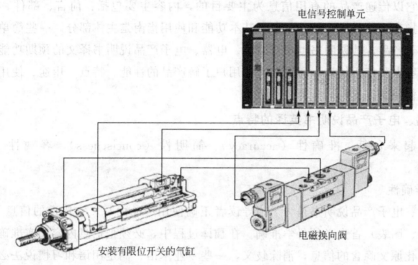

图12-3 一个现代电气气动控制回路的结构

- 电气信号控制部分通过电路的通断切换方向控制阀。
- 方向控制阀使活塞杆伸出和缩回。
- 活塞杆的位置由接近开关的电气信号部分检测。

Section VI of Translating Skills：家用电器、电子产品说明书的翻译

电器、电子产品说明书是科技文体的一种，也是机电专业学生将来在工作中经常接触到的，它以传递产品的有用信息为主要目的，内容主要包括：前言、部件、基本功能、使用指南、故障排除等。其中，基本功能和使用指南是主体部分。一些简单易用的电器、电子产品，其说明书也相对简略。电器、电子产品说明书译文的预期功能主要是提供商品特点和使用信息，通过译文让用户了解产品的性能、特点、用途、使用和保管等方面，促使其完成购买行为。

一、电器、电子产品说明书英译的特点

概括起来包括：准确性（accuracy）、简明性（conciseness）、客观性（objectivity）等。

1. 准确性

电器、电子产品说明书是为了指导读者正确使用产品而写，它传递的信息（例如：各种数据、图表）首先必须科学准确。在翻译过程中，必须把信息内容如实准确地翻译出来，显化原文隐含的信息，消除歧义。一些专业术语、固定用语和习惯说法必须表达得准确、地道，例如在翻译数码相机说明书时会遇到这样一些术语：镜头后盖（ear lens cap）、三脚架（tripod）、数码变焦（digital zoom）、快门帘幕（shutter curtain）、曝光不足（under exposure）、取景器（view finder）等，需按专业说法表达出来，不可任意生造。

2. 简明性

电器、电子产品说明书英译的简明性特点表现为以下几方面。

（1）内容条目简洁明了，步骤清晰，逻辑性强。例如，部件名称、操作界面等都配以示意图，再用箭头注明；操作步骤等用项目符号或编号依次标出；有些地方还把数据信息列成表格，简单明了，使人一目了然。

（2）常用缩略形式。例如：液晶显示（Liquid Crystal Display）常缩写成 LCD；发光二极管（Light Emitting Diode）常缩写成 LED；中央处理器（Central Processing Unit）常缩写成 CPU；自动对焦（Auto focus）常缩写成 AF；手动对焦（Manual focus）常缩写成 MF。

3. 客观性

电器、电子产品说明书将该产品的相关内容客观地呈现出来，引导读者按照一定的思维逻辑循序渐进，知道该做什么，怎么做，进而了解和正确使用该产品。这些内容带有描述说明的性质，客观而不带有感情色彩。例如，原文，紧急退出功能键可让使用者在电源故障时，以手动方式打开 CD 托盘；译文，The emergency-eject option allows

Unit 4　Pneumatics and Electropneumatics

the user to naturally open the CD tray during a power malfunction.

4. 准确性、简明性、客观性的共同体现

电器、电子产品说明书的英译具有准确、简明、客观等特点，这些特点共同体现在以下方面。

（1）广泛使用复合名词结构。在译文中复合名词结构代替各式后置定语，以求行文简洁、明了、客观，如：

原文，设备清单；译文，equipment check list（不用 the list of equipment check）。

原文，保修卡；译文，warranty card（不用 the card of warranty）。

有时候一些小标题常英译成动名词短语。

如：原文，测光模式；译文，metering modes.

译句常使用非人称名词化结构作主语，使句意更客观、简洁。如：

原文，由于使用了计算机，数据计算方面的问题得到了解决；译文，The use of computers has solved the problems in the area of calculating.

（2）普遍使用一般现在时。一般现在时可以用来表示不受时限的客观存在，包括客观真理、格言、科学事实及其他不受时限的事实。电器、电子产品说明书的主体部分就是进行"无时间性"（Timeless）的一般叙述，其译文普遍使用一般现在时，以体现出内容的客观性和形式的简明性。例如：

原文，本传真机与数码电话系统不兼容；译文，This facsimile machine is not compatible with digital telephone systems.

（3）常使用被动语态。电器、电子产品说明书英译的主要目的是说明相关产品（即受动者）的客观事实，其强调的是所叙述的事物本身，而并不需要过多地注意它的行为主体（即施动者）。这样的特点使得在其英译过程中大量使用被动语态，使译文客观简洁，而且可以使读者的注意力集中在受动者这一主要信息上。例如：原文，您可以在光盘中的电子使用手册中找到额外的信息；译文，Additional information can be found in the electronic user's manual which is located on the CD-ROM.

（4）广泛使用祈使句。电器、电子产品说明书很多地方都是指导使用者要做什么，不要做什么或该怎么做，所以其译文经常使用祈使句，谓语一般用动词原形，没有主语，译文的表述显得准确、客观而又简洁、明了。例如：

原文，请勿将 CF 卡存放在过热、多灰尘或潮湿的环境中，也不能存放在能产生静电荷或者电磁波的环境中；译文，Do not store CF cards in hot, dusty or humid places. Also avoid places prone to generate static charge or an electromagnetic field. （译文中出现了两个祈使句）

二、电器、电子产品说明书的英译技巧

电器、电子产品说明书很大篇幅是叙述使用方法和操作步骤，其语言平实，修辞手法单调，很少用到文学作品中常出现的比喻、拟人、夸张等修辞手法。因此，其译文也相应比较平实，英译时以直译为主，但是有时也要适当运用意译，正如王佐良在"词义·文体·翻译"一文中所写道的："一部好的译作总是既有直译又有意译的，凡能直译处

坚持直译，必须意译处则放手意译。"

1. 直译（literal translation）

在将中文的电器、电子产品说明书翻译成英文时，直译是最常用的技巧。周建人在为《外语教学与翻译》写的一篇题为"关于直译"的文章中曾写道："直译既不是'字典译法'，也不是死译、硬译，它是要求真正的意译，要求不失原文的语气与文情，确切地翻译过来的译法。"如：

原文：电池的使用寿命是10年。

译文：The battery's service life is 10 years.

2. 意译（free translation）

在英译过程中，将原文的一些词语或句子成分作适当调整，才能使译文更好地符合英语的表达习惯，这时就需要运用意译这一重要的翻译技巧，具体包括以下几方面。

（1）语序调整。汉英两种语言有不同的表达习惯，词和分句的顺序有时也不一样，如在表示时间地点时，汉语习惯先大后小，而英语则习惯先小后大。例如：

原文：监视器上没有影像。

译文：No pictures on the monitor.

（2）词类转换。翻译不是机械照搬，在作汉译英时，原文的某些词类应根据英语的表达习惯作适当转换才能使译文更自然、地道。汉语中的动词、名词、形容词、副词等在英译时都能转换成其他词类。例如：

原文：有些人赞成这个操作方法，而有些人反对。

译文：Some people are for the operation method, while some are against it.

分析：两个动词"赞成"和"反对"翻译成了介词"for"和"against"。

原文：这个机器帮助盲人行走。

译文：This machine helps the blind to walk。

分析：名词"盲人"翻译成了形容词。

原文：你会发现，这份示意图是十分有用的。

译文：You'll find this sketch map of great use。

分析：形容词"有用的"翻译成了名词"use"。

（3）词语增略。汉、英两种语言有不同的表达方式，为了使译文在语法、形式、意义等方面表达得完整贴切，常常在原文的基础上增补或省略必要的字、词、分句等。汉语中不用冠词，有时习惯省略一些名词、代词、连词或介词。在英译时，为了使译文在语法和结构上表述准确，需要增补这些成分。例如：

原文：封底功能启动时，您可以选择将作为标题出现在封底顶部的一条资讯。

译文：When the cover sheet function is turned on, you can select a message which will appear as a header at the top of the cover sheet.

分析：增译了冠词a和the，代词which。

词语的省略是指删去一些可有可无的词，或者有了反嫌累赘或违背译文语言习惯的

Unit 4 Pneumatics and Electropneumatics

词。它不是任意删减,而是使译文更简洁、精炼的一种翻译技巧。汉语中常用的量词,如"只、个、把、张、条"等,在英译时通常可以省略。

原文:务必将大齿轮上的两个突起部分与卷轴端子的凹槽牢固配合。

译文:Make sure the two protrusions on the large gear fit firmly into the slots in the end of the spool.

分析:翻译时省略了"个"。

三、机电产品说明书常见模式

机电产品说明书主要用于帮助使用者掌握该设备的操作方法,包括用途、产品规格、操作须知、维护及保养等。

1. 对用途的说明

These versatile wheeled tractors are reliable enough to tackle a variety of field jobs such as ploughing, harrowing, seed drilling, cultivating, harvesting, etc.

这种多用途轮式拖拉机适用于犁、耙、播种、耕作、收割等多种田间作业。

2. 对产品规格的描述

Processing range:6.5~2.4mm output:3.5~5t/shift

加工范围:6.5~2.4毫米 产量:3.5~5吨/班

3. 对设备特点的说明

Simple construction, easy operation and maintenance, and comparatively high productivity.

结构简单,操作容易,维修方便,生产率较高。

4. 故障排除

Trouble:The shaver does not work when the ON/OFF button is pressed.

Solution:Replace the batteries. If the shaver still does not work, see "Guarantee & Service".

问题:按下开/关按钮后剃须刀不工作。

解决方法:更换电池。如果剃须刀仍然不能工作,请参阅"保证及维修服务"。

5. 安全警示说明

Prevent the appliance and the wire from getting wet.

确保剃须刀与电线保持干燥。

Remove the batteries from the appliance if you are not going to use it for quite some time.

长时间不使用本设备,请取出电池。

6. 操作说明

Switch the appliance on by pressing the switch lock and pushing the ON/OFF button upwards.

按下开关键并把开/关按钮向上推,便可启动本设备。

第六节 翻译小练习

结合上面学习的关于家用电器、电子产品说明书的特点,认真阅读下面微波炉的英文使用说明书,并试着把它翻译出来。

Microwave Oven

1. Before use, please do read the Using Guidance.

2. Before use, the user must check whether the utensil is suitable for microwave oven; Be sure don't use the utensil which is not suitable for microwave oven to heat food in the microwave oven, so as to avoid abnormal occurrence such as fire and keep the microwave oven from damage.

3. The microwave oven is designed for food heating and cooking in family, can't be used for industry and commercial purposes.

4. If the outer housing, door or door seal of the microwave oven is damaged by collision or fall, please stop the use of it immediately.

5. Outer housing shutters can't be covered, in order to avoid the microwave oven be damaged by high temperature.

6. If smoke or spark appears during the use of the microwave oven, please keep the door off and immediately cut-off power supply.

7. The microwave oven can't be started without food in it. Empty running is harmful to the microwave oven, and may give rise to danger.

8. Without glass-turnplate and swivel piece, the microwave oven can't be used.

9. When the food is heated or cooked with the package of preservative film or plastic wrap, please keep the package is not fully closed so as to avoid burst.

10. The microwave oven can't be used to heat shelly fresh eggs or already-boiled eggs, because there may be a explosion when they are being heated or even after heated by the microwave oven.

11. Using microwave oven to fry and bake food, or to heat oily food, or for a long time heating, please make sure to monitor the situation of food cooking at any time in order to prevent fire.

12. When heating low moisture content food and food packaged by non-heatproof container, please use low heat so as to avoid the food or package catch fire.

13. When cooking soup or a big amount of food, the distance from the top of the food to the edge of the container should be more than 3.5cm. Otherwise, the food may boil and overflow.

14. When the ambient temperature is above 40℃, please don't use the microwave oven. High temperature operation may damage electrical parts.

15. When heating soup or drinks, please note that some liquid is free of bubbles when the temperature is above boiling point. Thellos may cause a sudden boiling phenomenon. Please wait some time before eat the food and stir it well, so as not to be scalded.

16. When cooking food, be sure not tightly seal or closed the container (the nipple should be unscrewed when the bottle is heated), feeding bottle or baby bottle should be stirred and shaken. Before feeding, check the temperature of the food in the bottle so as to prevent scald.

17. Cookbook only provides reference cooking time, but the cooking time is subject to the influence of raw and cooked degree each liked, the beginning temperature of the food, the amount, the size and the food container, ect. You can refer these factors to change the cooking time appropriately.

第六节翻译小练习答案

微波炉

1. 使用前，请阅读使用指南。

2. 使用前，用户必须检查所用器具是否适用于微波炉；确保不要使用不适合在微波炉中加热食物的器具，以免引起着火等异常现象，要保护微波炉以免受损。

3. 微波炉是专为家庭加热和烹调食物而设计的，不适用于工业和商业加热。

4. 如果外壳、门或微波炉门密封因碰撞或跌落损坏，请立即停止使用。

5. 外罩百叶窗严禁覆盖，以避免高温损坏微波炉。

6. 如果微波炉在使用过程中出现烟雾或火花，请保持门关闭，立即切断电源。

7. 微波炉不能在没有食物的情况下开启。空载运行会对微波炉有损害，从而增加危险性。

8. 没有玻璃转盘和旋转件，微波炉不能用。

9. 当食物用保鲜膜或塑料纸包加热时，请保持包装不完全关闭，以免爆裂。

10. 微波炉不能用于加热带壳的新鲜蛋或已煮熟的鸡蛋，因为在加热过程中可能会产生爆炸。

11. 用微波炉煎烤食物，加热油腻食物，或长时间加热食物时，请确保随时监测食品烹调情况以防止火灾的发生。

12. 当加热水分含量低的食品或用非耐热容器包装的食品时，请使用低温加热以避免食物或包装着火。

13. 当烹饪汤或大量的食物时，食物顶部与容器边缘的距离应大于3.5厘米。否则，食物可能煮沸溢出。

14. 当环境温度高于40℃，请不要使用微波炉。高温操作可能损坏电气部件。

15. 当加热汤或饮料时，请注意有些液体的温度是高于沸点的，这可能导致突然沸腾现象。请等一段时间再吃食物并搅拌均匀，以免烫伤。

16. 在烹调食物时，切勿密封或封闭容器（奶嘴奶瓶应拧下加热），喂食前，奶瓶或婴儿奶瓶应搅拌或摇动，检查瓶子里食物的温度以防止烫伤。

17. 烹饪手册提供的时间只供参考。实际烹饪时间受原材料、各人喜欢的生熟程度、烹饪的食物量、起始温度、容器尺寸与容量等的影响，你可以参考这些因素适当改变烹调时间。

Unit 5

Automatic Control Techniques

Lesson 13　The Introduction to Controls

Part 1　▶▶　Text

The subject of automatic controls is enormous, covering the control of variables such as temperature, pressure, flow, level, and speed.

The Controls Engineer needs to have various skills at his command—knowledge of mechanical engineering, electrical engineering, electronics and pneumatic systems, a working understanding of HVAC design and process applications and, increasingly today, an understanding of computers and digital communications.

The Need for Automatic Controls

There are three major reasons why process plant or buildings require automatic controls:

● Safety—the plant or process must be safe to operate. The more complex or dangerous the plant or process, the greater is the need for automatic controls and safeguard protocol.

● Stability—the plant or processes should work steadily, predictably and repeatably, without fluctuations or unplanned shutdowns.

● Accuracy—this is a primary requirement in factories and buildings to prevent spoilage, increase quality and production rates, and maintain comfort. These are the fundamentals of economic efficiency.

Other desirable benefits such as economy, speed, and reliability are also important, but it is against the three major parameters of safety, stability and accuracy that each control application will be measured.

Automatic Control Terminology

Specific terms are used within the controls industry, primary to avoid confusion. The same words and phrases come together in all aspects of controls, and when used correctly, their meaning is universal.

The simple manual system illustrated in Fig. 13-1 is used to introduce some standard terms used in control engineering.

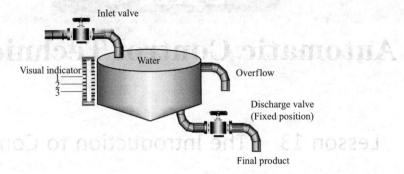

Fig. 13-1　A simple analogy of a control system

In the process example shown (Fig. 13-1), the operator manually varies the flow of water by opening or closing an inlet valve to ensure that:

1. The water level is not too high; or it will run to waste via the overflow.

2. The water level is not too low; or it will not cover the bottom of the tank.

The outcome of this is that the water runs out of the tank at a rate within a required range. If the water runs out at too high or too low a rate, the process it is feeding cannot operate properly.

At an initial stage, the outlet valve in the discharge pipe is fixed at a certain position.

The operator has marked three lines on the side of the tank to enable him to manipulate the water supply via the inlet valve. The 3 levels represent:

1. The lowest allowable water level to ensure the bottom of the tank is covered.

2. The highest allowable water level to ensure there is no discharge through the overflow.

3. The ideal level between 1 and 2.

The example (Fig. 13-1) demonstrates that:

1. The operator is aiming to maintain the water in the vessel between levels 1 and 2. The water level is called the Controlled Condition.

2. The controlled condition is achieved by controlling the flow of water through the valve in the inlet pipe. The flow is known as the Manipulated Variable, and the valve is referred to as the Controlled Device.

3. The water itself is known as the Control Agent.

4. By controlling the flow of water into the tank, the level of water in the tank is

altered. The change in water level is known as the Controlled Variable.

5. Once the water is in the tank it is known as the Controlled Medium.

6. The level of water trying to be maintained on the visual indicator is known as the Set Value (also known as the Set Point).

7. The water level can be maintained at any point between 1 and 2 on the visual indicator and still meet the control parameters such that the bottom of the tank is covered and there is no overflow. Any value within this range is known as the Desired Value.

8. Assume the level is strictly maintained at any point between 1 and 2. This is the water level at steady state conditions, referred to as the Control Value or Actual Value.

9. Note: With reference to (7) and (8) above, the ideal level of water to be maintained was at point 3. But if the actual level is at any point between 1 and 2, then that is still satisfactory. The difference between the Set Point and the Actual Value is known as Deviation.

10. If the inlet valve is closed to a new position, the water level will drop and the deviation will change. A sustained deviation is known as Offset.

Elements of automatic control (Fig. 13-2)

1. The operator's eye detects movement of the water against the marked scale indicator. His eye could be thought of as a Sensor (Fig. 13-2).

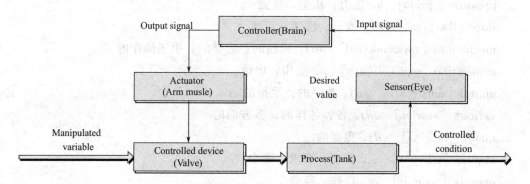

Fig. 13-2 A simple analogy of a control system

2. The eye (sensor) signals this information back to the brain, which notices a deviation. The brain could be thought of as a Controller.

3. The brain (controller) acts to send a signal to the arm muscle and hand, which could be thought of as an Actuator.

4. The arm muscle and hand (actuator) turn the valve, which could be thought of as a Controlled Device.

In simple terms the operators aim in example "A simple analogy of a control system" is to hold the water within the tank at a pre-defined level. Level 3 can be considered to be his target or Set Point.

The operator physically manipulates the level by adjusting the inlet valve (the control device). Within this operation it is necessary to take the operator's competence and concentration into account. Because of this, it is unlikely that the water level will be exactly at Level 3 at all times. Generally, it will be at a point above or below Level 3. The position or level at any particular moment is termed the Control Value or Actual Value.

The amount of error or difference between the Set Point and the Actual Value is termed deviation. When a deviation is constant, or steady state, it is termed Sustained Deviation or Offset.

Although the operator is manipulating the water level, the final aim is to generate a proper outcome, in this case, a required flow of water from the tank.

Part 2　New Words and Phrases

subject ['sʌbdʒekt] n. 主题, 科目, 主语, 国民
enormous [i'nɔːməs] adj. 庞大的, 巨大的, 凶暴的, 极恶的
covering ['kʌv(ə)riŋ] prep. 包括
temperature ['temprətʃə(r)] n. 温度
pressure ['preʃə] n. 压力; 压迫, 压强
flow [fləʊ] n. 流动; 流量; 涨潮, 泛滥
mechanical [mi'kænik(ə)l] adj. 机械的; 力学的; 手工操作的
application [æpli'keiʃ(ə)n] n. 应用; 申请
digital ['didʒit(ə)l] adj. 数字的; 手指的
various ['veəriəs] adj. 各种各样的; 多方面的
major ['meidʒə] adj. 重要的
require [ri'kwaiə] vt. 需要; 要求; 命令
operate ['ɒpəreit] vt. 操作; 经营
stability [stə'biliti] n. 稳定性; 坚定, 恒心
steadily ['stedili] adv. 稳定地; 稳固地
predictably [pri'diktəbli] adv. 可预言地
repeatably [ri'piːtəbli] n. 重复性
accuracy ['ækjʊrəsi] n. 精确度, 准确性
primary ['praim(ə)ri] adj. 主要的; 初级的; 基本的
fundamental [fʌndə'ment(ə)l] adj. 基本的
　　　　　　　　　　　　　　　　 n. 基本原理
parameter [pə'ræmitə] n. 参数; 系数
maintain [mein'tein] vt. 维持

Unit 5 Automatic Control Techniques

benefit ['benifit] *n.* 利益
　　　　　　　　　　vt. 有益于
specific [spə'sifik] *adj.* 特殊的
　　　　　　　　　　n. 特性；细节
terminology [,tɜ:mi'nɒlədʒi] *n.* 术语
confusion [kən'fju:ʒ(ə)n] *n.* 混乱；困惑
universal [,ju:ni'vɜ:s(ə)l] *adj.* 通用的
illustrate ['iləstreit] *vt.* 阐明，举例说明
analogy [ə'nælədʒi] *n.* 类比；类推
manually ['mænjʊəli] *adv.* 手动地
ensure [in'ʃʊə] *vt.* 确保；使安全
via ['vaiə] *prep.* 通过；经由
overflow [əʊvə'fləʊ] *n.* 充满，泛滥
tank [tæŋk] *n.* 坦克；水槽；池塘
properly ['prɒp(ə)li] *adv.* 适当地；正确地
initial [i'niʃəl] *adj.* 最初的；字首的
allowable [ə'laʊəbl] *adj.* 许可的；正当的
demonstrate ['demənstreit] *vt.* 证明；展示
satisfactory [,sætis'fækt(ə)ri] *adj.* 满意的；符合要求的
detect [di'tekt] *vt.* 发现；探测
slightly ['slaitli] *adv.* 些微地，轻微地
reinforce [,ri:in'fɔ:s] *vt.* 加强；强化
physically ['fizikəli] *adv.* 身体上，身体上地
adjust [ə'dʒʌst] *vt.* 调整，使……适合；校准
competence ['kɒmpit(ə)ns] *n.* 能力，胜任；权限；作证能力
concentration [,kɒns(ə)n'treiʃ(ə)n] *n.* 浓度；集中；浓缩；专心；集合
particular [pə'tikjələ(r)] *adj.* 特别的；详细的
term [tɜ:m] *n.* 术语；条款
　　　　　　vt. 把……叫做
amount [ə'maʊnt] *n.* 数量；总额
error ['erə] *n.* 误差；错误
constant ['kɒnst(ə)nt] *adj.* 不变的；恒定的
　　　　　　　　　　n. 常数；恒量
generate ['dʒenəreit] *vt.* 使形成；发生；生殖
at one's command　可自由支配；控制

be referred to as 被称为……
at steady state conditions 在稳态条件下
run out 耗尽；跑出
aim to 打算；目标在于……
be thought of as 被认为是，被看作
take...into account 把……考虑进去
sustained deviation 持续偏差
in this case 假若这样

Part 3 ▶▶ Technical Words and Phrases

variable ['veəriəb(ə)l] adj. 变量的
　　　　　　　　　　　　n. 变量，可变因素
HVAC abbr. 高压交流电（High Voltage Alternating Current）
　　　　　暖通空调（Heating Ventilation Air Conditioning）
fluctuation [flʌktʃʊ'eiʃ(ə)n] n. 起伏，波动
spoilage ['spɔilidʒ] n. 损坏，糟蹋；掠夺；损坏物
assume [ə'sju:m] vt. 承担；假定
deviation [di:vi'eiʃ(ə)n] n. 偏差；误差
offset ['ɒfset] n. 抵消，补偿
sensor ['sensə] n. 感应器
actuator ['æktjʊeitə] n. 促动器
automatic control 自动控制
mechanical engineering 机械工程；机械工程学
electrical engineering 电机工程，电气工程
pneumatic system 气压系统；空气系统，通气系统
HVAC design 暖通设计
mechanical engineering 机械工程；机械工程学
electrical engineering 电机工程，电气工程
process building 工艺过程厂房
process plant 制炼厂
automatic control 自动控制
safeguard protocol 维护协议
production rate 生产率
control application 控制应用，控制应用程序
economic efficiency 经济效益
manual system 人工系统；手动系统

control engineering   控制工程
inlet valve   进气阀；进给阀
outlet valve   排出阀
discharge pipe   排放管；卸料管；散热管
controlled condition   受控条件，受控状态
manipulated variable   操纵量，被控变量
controlled device   控制装置，控制器
control agent   控制媒体，调节体
controlled medium   控制介质，控制媒体
visual indicator   指示剂；视觉指示器
set value   设定值，给定值；凝固值；蝶点
set point   设定值，调整点；凝结点
control parameter   控制参数
desired value   期望值，希望值；需要值
control value   控制值
actual value   实际价值；实际指标
marked scale indicator   显著的规模指标
control valve   控制阀
actual valve   实际价值；实际指标

Part 4 ▶▶ Translations

1. The subject of automatic controls is enormous, covering the control of variables such as temperature, pressure, flow, level, and speed.
自动控制技术意义广泛，涵盖变量的控制，如温度、压力、流量、液位和速度。

2. The more complex or dangerous the plant or process, the greater is the need for automatic controls and safeguard protocol.
越危险或越复杂的工厂，越需要自动控制和安全维护协议。

3. Other desirable benefits such as economy, speed, and reliability are also important, but it is against the three major parameters of safety, stability and accuracy that each control application will be measured.
其他必要的因素，如经济、速度和可靠性也很重要。但是却会和三个主要因素（安全、稳定和准确）相悖，而它们又是必须考量的应用。

4. The water level is not too high; or it will run to waste via the overflow.
水位低，会溢出，产生浪费。

5. At an initial stage, the outlet valve in the discharge pipe is fixed at a certain position.

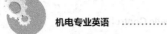

在初始阶段，放电管的出口阀固定在一定的位置。

6. The flow is known as the Manipulated Variable, and the valve is referred to as the Controlled Device.

通过控制进给阀改成的水流实现控制条件。流量即可控变量，阀即控制设备。

7. The difference between the Set Point and the Actual Value is known as Deviation.

设置点和实际值间的差别即误差。

8. The eye (sensor) signals this information back to the brain, which notices a deviation. The brain could be thought of as a Controller.

眼睛（感应器）将信息传给大脑，大脑测算误差，大脑这时就是控制器。

9. Within this operation it is necessary to take the operator's competence and concentration into account.

实际上，操作员通过调整进给阀（控制装置）进行操作。操作过程中，需要考虑操作员的能力和注意力。

10. The amount of error or difference between the Set Point and the Actual Value is termed deviation.

设定点和实际值间的差别即误差。若误差恒定，即为持续误差或持续偏移。

Part 5　Reference Version

第十三课　自动控制技术简介

自动控制技术意义广泛，涵盖诸如温度、压力、流量、液位和速度等变量的控制。

控制工程师需要掌握各门学科——机械工程、电气工程、电子和气动系统，并且擅长暖通空调设计和过程应用。还包括现如今越来越需要的计算机和数字通信。

自动控制技术的意义

● 安全——工厂操作过程必须是安全的。越危险或越复杂的工厂，越需要自动控制和安全维护协议。

● 稳定——工厂要稳定运行，且具有前瞻性和循环性，不能出现波动或意外关闭。

● 准确——避免损坏、提高质量和生产率、保证舒适度是对工厂和建筑物的基本要求。这些都是经济效益的基本要素。

其他必要的因素，如经济、速度和可靠性也很重要。但是却会和三个主要因素（安全、稳定和准确）相悖，而它们又是必须考量的应用。

自动控制术语

掌握自动控制领域的专用术语是避免混淆的基本要求。相同的单词和短语用于描述控制领域各个方面，如果正确使用，其意思是通用的。

如图 13-1 所示的简单的手控系统，可以说明控制工程中的标准条款。

图 13-1 所示的流程示例中，操作员通过开关阀门来手动改变流量，确保如下两点：

1. 水位不能过高，否则会通过溢出口流出导致浪费。
2. 水位不能太低，否则不能覆盖水箱的底部。

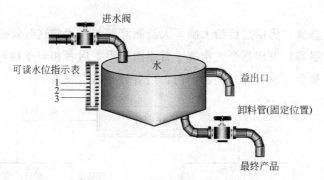

图 13-1　水位模拟控制系统

这一要求的结果是水从水箱流出的速率必须在允许的范围内。如果水以过高或过低的速率耗尽，进给过程不能正常运转。

在初始阶段，出水管路中的出水阀开度固定在一定的位置上。

操作员已经在水箱的一侧标记了三条线，这样可以使得他能够通过操作进水阀供水。三条线分别代表：

1. 允许的最低水位，确保水箱底部被覆盖。
2. 允许的最高水位，确保水不会通过溢出口溢出。
3. 理想水位线是介于 1 和 2 的中间那条。

图 13-1 表明：

1. 操作者意在维持容器中的水位在 1 和 2 之间。水位即为被控对象。
2. 通过控制进水管道中的进给阀实现被控对象的控制。流量被称为操纵变量，阀被称为控制设备。
3. 水本身被称为控制要素。
4. 通过控制进入水箱的水流，水箱的水位得以改变。水位的变化为可控变量。
5. 一旦水流入水箱中，它就是控制介质。
6. 水位如果试图在可视指示器上保持固定位置的话，这一位置就可称为设定值（也称为设置点）。
7. 水位可以保持在可视指示器的 1 点和 2 点之间的任何一个位置上，只要水箱底部被水覆盖或者水没有溢出就符合控制参数。该范围内的任何值为测量值。
8. 假设水位被严格保持在 1 和 2 之间某处，这种在稳定状态下的水位，被称为实际值。
9. 注意：根据 7. 和 8.，理想水位应保持在第 3 位置。但如果实际值在 1 和 2 间的某一点上，仍旧保持稳定状态，那么设定点和实际值间的差别即偏差。

10. 若进给阀被关闭到某一新位置，水位将下降，偏差将发生变化。持续的偏差被称为偏移量。

自动控制要素（见图 13-2）

1. 操作员通过比对刻度指示器，用眼睛检测水位变化。他的眼睛相当于传感器（图 13-2）。

2. 眼睛（传感器）将信息传给大脑，大脑测算偏差。大脑这时就是控制器。

3. 大脑（控制器）发出指令，指挥手臂肌肉和手，后者相当于执行器。

4. 手臂肌肉和手（执行器）控制阀门，后者就是被控设备。

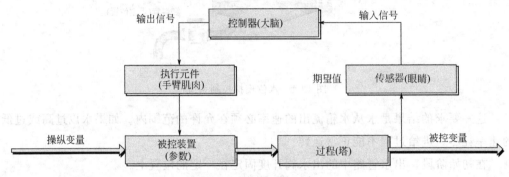

图 13-2 控制系统要素

以上这个简单的"单值控制系统"例子，操作员的目的是将水位保持在预定位置上。水位 3 即目标值或设定值。

实际上，操作员通过调整进给阀（控制装置）进行操作。操作过程中，需要考验操作员的持续力和注意力。所以，水位不会持续精确地停留在位置 3，会出现过高或过低的情况。在任何特定时刻的水位被定义为控制值或实际值。

设定点和实际值间的差别即偏差。若偏差持续或者固定，即被称为固有偏差或偏移量。

虽然操作员控制水位，但最终目标是产生合适的输出量，在这个例子中，是水箱的出水量。

Lesson 14 The Programmable Logical Controller Techique

Part 1 ▶▶ Text

What is a PLC?

A PLC (ie Programmable Logic Controller) is a device that was invented to replace the necessary sequential replay circuits for machine control. The PLC works by looking at its inputs and depending upon their state, turning on/off its outputs. The user enters a program, usually via software, that gives the desired results.

The Guts Inside

The PLC mainly consists of a CPU, memory areas, and appropriate circuits to receive input/output data. We can actually consider the PLC to be a box full of hundreds or thousands of separate relays, counters, timers and data storage locations (Fig. 14-1). Do these counters, timers, etc. Really exist? No, they don't "physically" exist but rather they are simulated and can be considered software counters, timers, etc. These internal relays are simulated through bit locations in registers. (more on that later)

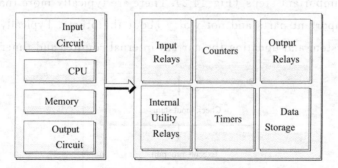

Fig. 14-1 The guts inside

What Does Each Part Do?

INPUT RELAYS—These are connected to the outside world. They physically exist and receive signals from switches, sensors, etc. Typically they are not relays but rather they are transistors.

　INTERNAL UTILITY RELAYS—These do not receive signals from the outside world nor do they physically exist. They are simulated relays and are what enables a PLC to eliminate external relays. There are also some special relays that are dedicated to performing only one task. Some are always on while some are always off. Some are on only once during power-on and are typically used for initializing data that was stored.

COUNTERS—There again do not physically exist. They are simulated counters and they can be programmed to count pulses. Typically these counters can count up, down or both up and down. Since they are simulated they are limited in their counting speed. Some manufacturers also include high-speed counters that are hardware based. We can think of these as physically existing. Most times these counters can count up, down, or up and down.

OUTPUT RELAYS—These are connected to the outside world. They physically exist and send on/off signals to solenoids, lights, etc. They can be transistors, relays, or triacs depending upon the model chosen.

DATA STORAGE—Typically there are registers assigned to simply store data. They are usually used as temporary storage for math or data manipulation. They can also typically be used to store data when power is removed from the PLC. Upon power-up they will still have the same contents as before power was removed. Very convenient and necessary!

TIMERS—These also do not physically exist. They come in many varieties and increments. The most common type is an on-delay type. Others include off-delay and both retentive and non-retentive types. Increments vary from 1 ms through 1 s.

How It Works?

A PLC works by continually scanning a program. We can think of this scan cycle as consisting of 3 important steps (Fig. 14-2). There are typically more than 3 but we can focus on the important parts and not worry about the others. Typically the others are checking the system and updating the current internal counter and timer values.

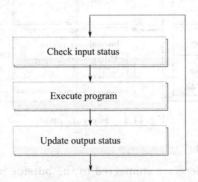

Fig. 14-2 PLC work steps

CHECK INPUT STATUS—First, the PLC takes a look at each input to determine if it is on or off. In other words, is the sensor connected to the first input on? How about the second input? How about the third... It records this data into its memory to be used during the next step.

EXECUTE PROGRAM—Next the PLC executes your program one instruction at a

Unit 5 Automatic Control Techniques

time. Maybe your program said that if the first input was on then it should turn on the first output. Since it already knows which inputs are on/off from the previous step it will be able to decide whether the first output should be turned on based on the state of the first input. It will store the execution results for use later during the next step.

UPDATE OUTDATE STATUS—Finally the PLC updates the status of the outputs. It updates the outputs based on the inputs during the first step and the results of executing your program during the second step. Based on the example in step 2 it would now turn on the first output because the first input was on and your program said to turn on the first output when this condition is true.

After the third step the PLC goes back to step one and repeats the steps continuously. One scan time is defined as the time it takes to execute the 3 steps listed above.

Part 2 ▶▶ New Words and Phrases

sequential [si'kwenʃ(ə)l] *adj.* 连续的；相继的；有顺序的
relay ['riːlei] *n.* 继电器；接替，接替人员；
simulate ['simjʊleit] *vt.* 模仿；假装；冒充
internal [in'tɜːn(ə)l] *adj.* 内部的；内在的；国内的
eliminate [i'limineit] *vt.* 消除；排除
typically ['tipikəli] *adv.* 代表性地；作为特色地
manufacturer [ˌmænjʊ'fæktʃ(ə)rə(r)] *n.* 制造商；厂商
variety [və'raiəti] *n.* 多样；种类；杂耍；变化，多样化
execute ['eksikjuːt] *vt.* 实行；执行；处死
execution [ˌeksi'kjuːʃ(ə)n] *n.* 执行，实行；完成；死刑
previous ['priːviəs] *adj.* 以前的；早先的；过早的
depend upon 依赖；取决于
consist of 由……组成；由……构成；包括
a box full of 满满的
be dedicated to 奉献；从事于；献身于
output relay 输出继电器

Part 3 ▶▶ Technical Words and Phrases

programmable [ˌprəʊ'græməbl] *adj.* 可编程的；可设计的
logical ['lɒdʒik(ə)l] *adj.* 合逻辑的，合理的；逻辑学的
circuit ['sɜːkit] *n.* 电路，回路；巡回；一圈；环道

input ['input] n. 投入；输入电路
output ['autput] n. 输出，输出量
program ['prəugræm] n. 程序；计划；大纲
gut [gʌt] n. 内脏；勇气；直觉
CPU 中央处理器（Central Processing Unit）
register ['redʒistə] vt. 登记；记录
n. 登记
timer ['taimə] n. 定时器；计时器；
switch [switʃ] n. 开关
transistor [træn'zistə] n. 晶体管（收音机）
pulse [pʌls] n. 脉冲；脉搏
increment ['inkrim(ə)nt] n. 增加；盈余
retentive [ri'tentiv] adj. 保持的；记性好的
non-retentive 非顽磁的；软磁的
solenoid ['səulənɔid] n. 螺线管；螺线形电导管
triac ['traiæk] 三端双向晶闸管开关元件
manipulation [mə, nipjʊ'leɪ(ə)n] n. 操纵；处理
scan [skæn] vt. 扫描
programmable logic controller 可编程序逻辑控制器
data storage location 数据存储位置
software counter 软件计数器
utility relay 继电保护
external relay 外部继电器
simulated counter 模拟计数器

Part 4 ▶▶ Translations

1. A PLC (ie Programmable Logic Controller) is a device that was invented to replace the necessary sequential replay circuits for machine control.

PLC（可编程序逻辑控制器）是很有特色的设备，它可以为机床控制更换必要的继电器电路。

2. The PLC mainly consists of a CPU, memory areas, and appropriate circuits to receive input/output data.

PLC 主要由 CPU、内存区域和适当接收输入/输出数据的电路组成。

3. Some manufacturers also include high-speed counters that are hardware based.

一些制造商认可基于硬件的高速计数器的效力，所以也可以认为计数器是存在的。

4. They can also typically be used to store data when power is removed from the

PLC. Upon power-up they will still have the same contents as before power was removed. Very convenient and necessary!

如果 PLC 的电源被切断，也可以存储数据，一旦恢复电源，内容也可以恢复。

5. There are typically more than 3 but we can focus on the important parts and not worry about the others.

通常多于三个步骤，但我们重视重要步骤而不是次要步骤。

6. It records this data into its memory to be used during the next step.

要通过记忆这些数据才决定下一步骤。

7. Since it already knows which inputs are on/off from the previous step it will be able to decide whether the first output should be turned on based on the state of the first input.

因为已经记忆了上一步输入开或关，就可以决定第一次输出是否可以开始。

8. After the third step the PLC goes back to step one and repeats the steps continuously. One scan time is defined as the time it takes to execute the 3 steps listed above.

第三步骤之后，PLC 回到第一步，重复这些步骤。

Part 5 Reference Version

第十四课 可编程序逻辑控制器技术

什么是 PLC？

PLC（可编程序逻辑控制器）是一种可以替换控制机器所必需的继电器电路的设备。PLC 通过检测输入设备的状态，运行程序，从而打开或关闭输出。使用者通过编程软件输入程序，从而给出期望的结果。

内部关键组成

PLC 主要由 CPU、存储区和特定输入/输出数据的电路组成。PLC 内部包含有诸多独立继电器、计数器、计时器和数据存储区（图 14-1）。但是这些计数器、计时器等实际上是不存在的。它们不是真正地存在，而是模拟出来的，可以被认为是软计数器、软计时器等。这些内部继电器是通过寄存器中的位单元模拟出来的。（稍后详细介绍）

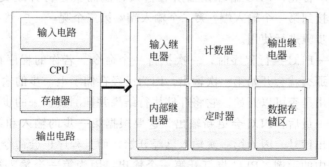

图 14-1 PLC 内部元件

PLC 内部元件的作用

输入继电器——与外部信号联系。真实存在并接收来自开关、传感器等的信号。通常它们不是继电器而是晶体管。

内部继电器——不接收外来信号,不真正存在,是模拟继电器,可以让PLC省掉输出继电器。有些继电器只执行一项任务;有些处于常开的状态,有些处于常关的状态;有些只在接通电源瞬间触发一次,用于初始化数据。

计数器——没有物理存在,是模拟计数器和可编程计数脉冲。通常可加可减,或可加减。由于是模拟的,计算速度很有限。一些制造商将基于硬件的高速计数器也涵盖在内,所以也可以认为计数器是存在的。大多数时候,这些计数器可加可减,或可加减。

输出继电器——与外部信息相连,真正存在,能够给螺线管和灯等发送信号。它们可以根据需要选择晶体管、继电器或三端双向晶闸管开关元件等类型。

数据存储区——通常,寄存器会简单地用于存储数据,通常也会临时存储数学信息或数据操作。如果PLC的电源被切断,也可以存储数据,一旦恢复电源,内容也可以恢复。非常方便和必要!

计时器——同样物理上不存在。类型和增量颇多。最常见的是接通延时类型,还有包括延迟断开的类型、保持型和非保持的类型。增量从1ms到1s内变化。

PLC是如何运行的?

PLC连续不断地扫描程序,扫描循环由三个重要步骤(图14-2)组成。通常多于三个步骤,但我们重视重要步骤而不是次要步骤。次要步骤一般是要检查系统和更新当前的内部计数器和定时器值。

图14-2 PLC工作过程

输入扫描——首先,PLC循环扫描所有输入,监测其是打开或关闭。换句话说,传感器是否应连接到第一个输入端?第二个输入端?第三个呢……?然后通过存储这些数据,以备下一步骤进行使用。

执行程序——PLC执行程序时,一次执行一个命令。若你的程序在第一次输入数据时决定打开,那么就会开始第一次输出。因为已经存储了上一步的输入开或关,就可以决定第一次输出是否可以开始。它将存储执行结果并推测接下来的步骤。

输出刷新——最后,PLC会更新输出。这要根据第一步的输入和第二步执行程序的结果来刷新输出信号。因为第一次输入已经开始,而且程序也会命令第一次输出的开始。根据步骤2中的示例,PLC将打开第一个输出,因为第一个输入条件为ON并且程序决定当这个条件为ON,则打开第一个输出。

第三步骤之后,PLC回到第一步,重复这些步骤。扫描时间定义为执行以上三个步骤的时间之和。

Lesson 15 Open Loop Control System and Closed Loop Control

 Text

An Open Loop Control System

Open loop control simply means there is no direct feedback from the controlled condition; in other words, no information is sent back from the process or system under control to advise the controller that corrective action is required. The heating system shown in Fig. 15-1 demonstrates this by using a sensor outside of the room being heated. The system shown in Fig. 15-1 is not an example of a practical heating control system; it is simply being used to depict the principle of open loop control.

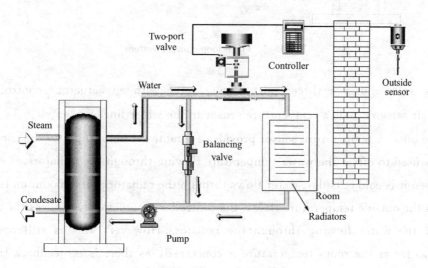

Fig. 15-1 Open loop control

The system consists of a proportional controller with an outside sensor sensing ambient air temperature. The controller might be set with a fairly large proportional band, such that at an ambient temperature of $-1\,℃$ the valve is full open, and at an ambient of $19\,℃$ the valve is fully closed. As the ambient temperature will have an effect on the heat loss from the building, it is hoped that the room temperature will be controlled.

However, there is no feedback regarding the room temperature and heating due to other factors. In mild weather, although the flow of water is being controlled, other factors, such as high solar gain, might cause the room to overheat. In other words, open control tends only to provide a coarse control of the application.

Fig. 15-2 depicts a slightly more sophisticated control system with two sensors.

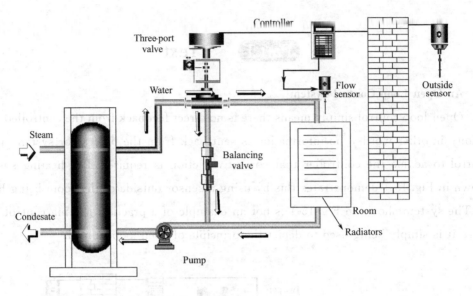

Fig. 15-2　Open loop control system

The system uses a three—port mixing valve with an actuator, controller and outside air sensor, plus a temperature sensor in the water line.

The outside temperature sensor provides a remote set point input to the controller, which is used to offset the water temperature flowing through the radiators.

When it is cold outside, water flows through the radiator at its maximum temperature. As the outside temperature rises, the controller automatically reduces the temperature of the water flowing through the radiators. However, this is still open loop control as far as the room temperature is concerned, as there is no feedback from the building or space being heated. If radiators are oversized or design errors have occurred, overheating will still occur.

A Closed Loop Control System

Quite simply, a closed loop control requires feedback; information sent back direct from the process or system. Using the simple heating system shown in Fig. 15-3, the addition of an internal space temperature sensor will detect the room temperature and provide closed loop control with respect to the room.

In Fig. 15-3, the valve and actuator are controlled via a space temperature sensor in the room, providing feedback from the actual room temperature.

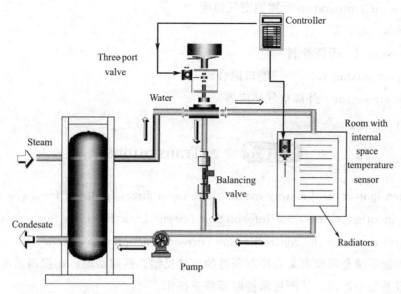

Fig. 15-3 Closed loop control system

Part 2 New Words and Phrases

feedback ['fiːdbæk] n. 反馈
depict [di'pikt] vt. 描述；描画
proportional [prə'pɔːʃən(ə)l] adj. 成比例的；相称的
solar ['səʊlər] adj. 太阳的；日光的
sophisticated [sə'fɪstɪkeɪtɪd] adj. 见多识广的；老练的；见过世面的；复杂的
maximum ['mæksɪməm] n. 极大，最大限度
addition [ə'dɪʃ(ə)n] n. 添加；加法；增加物
in other words 换句话说
under control 处于控制之下；情况正常
mix…with… （使）与……混合；（使）与……交融

Part 3 Technical Words and Phrases

loop [luːp] n. 环；圈
ambient ['æmbiənt] adj. 周围的；外界的
actuator ['æktʃʊeɪtə] n. 促动器
radiator ['reɪdieɪtə(r)] n. 散热器；暖气片；辐射体
open loop control system 开环控制系统
closed loop control system 闭环控制系统

ambient air temperature　周围空气温度

heat loss　热损失

coarse control　粗调控制

three-port mixing valve　三端口混合阀

outside air sensor　外部空气感应器

Part 4　Translations

1. Open loop control simply means there is no direct feedback from the controlled condition; in other words, no information is sent back from the process or system under control to advise the controller that corrective action is required.

开环控制系统是指没有来自控制条件的直接反馈。换句话说，在控制系统下没有信息从程序或系统中返回，从而提示控制器修正操作。

2. The system consists of a proportional controller with an outside sensor sensing ambient air temperature.

该系统带有可以感应周围温度的控制器。

3. The controller might be set with a fairly large proportional band, such that at an ambient temperature of －1℃ the valve is full open, and at an ambient of 19℃ the valve is fully closed.

该控制器自身调控范围很大，在环境－1℃温度条件下阀门会完全打开，而在19℃温度条件下阀门则会完全闭合。

4. In mild weather, although the flow of water is being controlled, other factors, such as high solar gain, might cause the room to overheat.

在温和的天气里，水流虽被控制，但其他因素，如太阳能，可能会导致温度过高。

5. The system uses a three-port mixing valve with an actuator, controller and outside air sensor, plus a temperature sensor in the water line.

该系统是三端口混合阀，带制动器、控制器和外部空气感应器、水位温度感应器。

6. As the outside temperature rises, the controller automatically reduces the temperature of the water flowing through the radiators.

随着温度的上升，控制器自动降低流经散热器的水温。

7. If radiators are oversized or design errors have occurred, overheating will still occur.

如果散热器超大或设计出现错误，仍会有过热的情况。

8. Quite simply, a closed loop control requires feedback; information sent back direct from the process or system.

闭环控制系统需要反馈，直接从流程或系统传回来的信息。

Part 5 ▶▶ Reference Version

第十五课 开环控制系统和闭环控制系统

开环控制系统

开环控制系统简单来讲是指没有来自被控变量的直接反馈。换句话说,在控制系统下没有信息从过程或系统中返回,从而提示控制器修正操作。图15-1的供热系统通过使用一个安装在被加热的房间外的传感器来供热。该系统图不是一个实际热控制系统,只是用来描述开环控制原理。

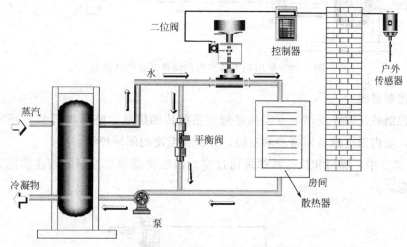

图15-1 供热系统开环控制示意图

该系统带有一个可以感应周围空气温度的传感器的控制器。该控制器自身调控范围很大,在环境-1℃温度条件下阀门会完全打开,而在19℃温度条件下阀门则会完全闭合。环境温度会引起热损失,这是不希望出现的状况。

然而,由于某些因素,环境温度和热量没有任何反馈。在温和的天气里,水流虽被控制,但其他因素,如太阳能,可能会导致温度过高。换句话说,开环控制系统往往只提供低级的应用控制。

图15-2描述了一个不太复杂的有两个传感器的控制系统。

该系统是三端口混合阀,带执行器、控制器和外部空气传感器加上一个水位温度感应器。

外部温度感应器提供了一套远程设定点给控制器,用于抵消流经散热器的水温损失。

若外部温度很低,水流经散热器时是它的最高温度。随着温度的上升,控制器自动降低流经散热器的水温。然而,就室温而言这仍然是开环控制系统,因为没有来自受热建筑或空间的反馈。如果散热器超大或设计出现错误,仍会有过热的情况。

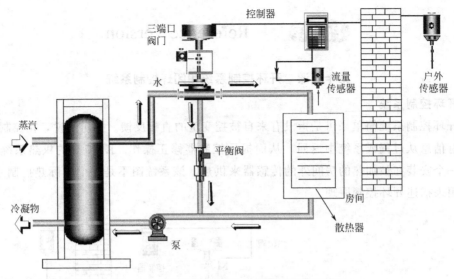

图 15-2 使用两个传感器的供热系统开环控制

闭环控制系统

闭环控制系统需要反馈，直接从过程或系统传回信息。使用如图 15-3 所示的简单加热系统，室内温度感应器将检测室温，并提供温度的闭环控制。

在图 15-3 中，阀门和执行器都是通过室内温度传感器来控制的，在实际室温中得到反馈信息。

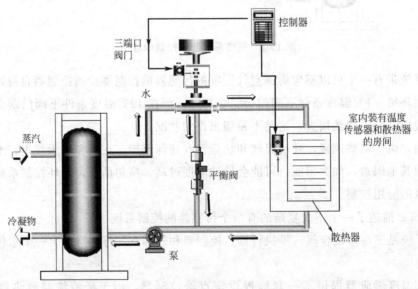

图 15-3 供热系统闭环控制示意图

Section Ⅶ of Translating Skills: 科普文章的翻译

科普就是科学技术普及的简称。人类的科学和技术活动，包括两个方面：一是科学技术的研究与开发；二是科学技术的传播与人才培养。因此科普是科技工作的重要组成部分。科普文就是把人类研究开发的科学知识、科学方法，以及融化于其中的科学思想和科学精神，通过多种方法，多种途径传播到社会的方方面面。科普文章的文体特征主要有下面几个方面：

（1）文体质朴，语言浓缩；
（2）注重客观性，描述平易、精确，不带个人感情色彩；
（3）结构严密，前言后语关系明确，主次分明，不允许有歧义现象；
（4）逻辑性强，表达明晰和流畅。

科普文章也有其句法上的特征：

（1）在人称使用上，强调文章内容的客观性，汉语中常使用"笔者"、"我们"、"本文"而不使用"我"来充当得出某一结论的主语；英语中则常用 The author thinks that…/The authors have investigated…/This paper illustrates…等句式。

（2）时态变化不明显。汉语里的动词无时态变化，动词所发生的时间往往是通过时间状语表示；英语中叙述客观事实时几乎千篇一律使用现在时。

（3）在语态方面，汉语中主动语态使用较多，而英语中，被动语态使用较多。

（4）汉语和英语的科技实用文中，句子结构都较为复杂。

下面我们来看几个例子。

实例一：

原文：The event-driven nature of Visual Basic allows you to build your application in stages and test it at each stage. You can build one procedure, or part of a procedure, at a time and try it until it works as desired. This minimizes errors and gives you, the programmer, confidence as your application takes shape.

第一句，The event-driven nature of Visual Basic allows you to build your application in stages and test it at each stage. 采用"断句"的翻译技巧，把主语单独列出，达到强调效果。两个动词所接的宾语其实是一致的，可以采用省略的方法。

第二句，You can build one procedure, or part of a procedure, at a time and try it until it works as desired. "until it works as desired"译成"直到它如预期的那样工作"固然不错，但是拘泥于原文，有些"洋"味，改译成"直到满足预期效果"，这样符合汉语表达习惯。

第三句，This minimizes errors and gives you, the programmer, confidence as your application takes shape. 对于"takes shape"，词典上解释如下："成形，形成，具

体化，有显著发展"，好像每个词义用上去都不合适，其实，形成应用程序，在程序员眼里，就是开发应用程序，为了让译文更符合汉语表达习惯，可以不把"as your application takes shape"当成时间状语来翻译，而是作为"give"宾语补足语来翻译，即"gives you confidence to develop your application"，采用了"转换成分：词性和句式的转换"的翻译技巧。

译文：VB的事件驱动特性，可使你分阶段创建和测试应用程序。你可每次创建一个过程，或者过程的一部分，然后测试它，直到满足预期效果。这就把错误减到最少，而且给你这个程序员以信心来开发应用程序。

实例二：

原文：As you progress in your programming skills, always remember to take this sequential approach to building a Visual Basic application. Build a little, test a little, modify a little and test again. You'll quickly have a completed application. This ability to quickly build something and try it makes working with Visual Basic fun-not a quality found in some programming environments! Now, we'll look at each step in the application development process

第一句，As you progress in your programming skills, always remember to take this sequential approach to building a Visual Basic application. "As you progress in your programming skills"译成"当你在编程技巧方面取得进步"固然不错，采用"转换成分"技巧，改译成"随着你编程技能的进步"更简洁流畅。

第三句，"You'll quickly have a completed application."译成"你将很快编好一个应用程序。"更符合汉语表达习惯。

第四句，"This ability to quickly build something and try it makes working with Visual Basic fun." "This ability to quickly build something and try it"同样采用省略合译的方法。采用"断句"技巧，前后两部分分译，读起来节奏要舒缓些。

第五句，"Now, we'll look at each step in the application development process."译时采用了"减词译法"的技巧，因为"开发"本身就是一个"过程"，所以不翻译"process"这个词，让译文更简洁。

译文：随着你编程技能的进步，始终记住创建VB应用程序的方法。创建一点，测试一点，修改一点，然后再测试。你将很快编好一个应用程序。这个快速创建和测试的能力，使得VB编程充满乐趣，这是某些编程环境不具备的。现在，让我们来看看开发应用程序的每个步骤。

第七节 翻译小练习

将下列短文译为汉语。

1. The champions of hearing, by any standard, are the bats. Bat sounds long went undetected by man because they are pitched two to three octaves（八度音阶）above what we can hear. But to a number of bats flying around on a calm, still summer eve-

ning—and to the unfortunate moths (蛾) that can hear them and must try to avoid them—the evening is anything but calm. It is a madhouse of constant screaming. Each bat sends out a series of screams in short pulses, each lasting less than a hundredth of a second.

2. If ignorance about the nature of pain is widespread, ignorance about the way pain-killing drugs work is even more so. What is not generally understood is that many of the boasted pain-killing drugs conceal the pain without correcting the underlying condition. The abuse of pain killing drugs will deaden the mechanism in the body that alerts the brain to the fact that something may be wrong. The body can pay a high price for suppression of pain without regard to its basic cause.

第七节翻译小练习答案

1. 不管用哪一种标准来衡量，听力的冠军都是蝙蝠。蝙蝠发出的声音在长时间内不为人类所觉察到，因为它们发出的声音要比我们所能听见的高2到3个八度。但是对于在安宁而又寂静的夏天的傍晚到处飞翔的许多蝙蝠——以及对于那些能听见它们声音而必须设法避开它们的不幸的蛾来说，这个夜晚一点儿也不宁静。那简直是一个不断在发出尖叫声的疯人院。每一只蝙蝠发出一连串短脉冲的尖叫声，每一声的持续时间还不到百分之一秒。

2. 如果说人们对疼痛的本质普遍不甚了解，那么他们对止痛药是如何起作用的就更是一无所知了。人们所知甚少的是许多被吹嘘为止痛药的药物在止痛时并没有消除产生疼痛的根源。滥用止痛药会使人体中向脑子警告有什么地方可能出了毛病的机制变麻木。光是镇痛而不考虑疼痛的根本原因，会使人体付出高昂的代价。

Lesson 16 "The Sage of PID"——An Educational Tale for Those Who Would Understand the Concept of PID (1)

Part 1  Text

Once upon a time, long ago in a far distant land, there was a council engineer called Emil. He was in charge of the water supply for a fabulous city and was about to retire because of his advanced age.

Of course he had organised his pension, social security and so on. In addition he had been foresighted enough to make an agreement with the council that his successor should be one of his sons: Peter, Ivan, David. But it had not been decided which one. Here is their tale:

1. Job description

Emil, the father, had been responsible for keeping a constant level (actually the aim was to achieve a constant pressure to the consumers) in the council's water tower independent of changes in consumption. This was done by opening or closing the three valves in the water supply line to the tower, depending on the consumption.

It should be noted that this was no easy job, since the consumption varied a lot. The sons had visited Emil at the tower now and then, but they had never taken the trouble to watch Emil's techniques. They had also followed the development of medieval control technology and considered themselves fully competent to take over.

Just before Emil was to retire, Osquar, a younger relative who had studied control engineering, visited Emil. They discussed control engineering at length and with great vision and agreed that a level control could be built up several different ways, using:

Manual

Electrical

Fluid power

Mechanical

Hydraulic or Pneumatic control

2. Three requirements

In any case the control had to comply with three requirements:

• When the level changes because of an "interference" (change in the consumption), the controller should take action against it as quickly and efficiently as possible to restore the required level.

• The return to the required level has to take place with as few deviations as possible and preferably not periodic.

• The controller has to keep the level within the necessary limits to achieve an even pressure in the supply lines.

3. The controllers are tested

Emil's retirement day was getting closer, but so far none of the sons had come up with a method or equipment which complied with the requirements.

The local council discussed the situation and agreed upon making a test where the sons should each show their skills with the father as a supervisor. This was an empirical method to find the best successor.

4. Ivan is oscillating

Ivan starts off with a methodical approach. When he sees that the water level has dropped 100mm below the ideal level-the set point-he realises that there is a rise in the water consumption. Therefore he starts to open the inlet valve slowly and evenly. After a while Ivan recognises that the reduction of the level is getting slower and at the end the level starts to rise again. But, still he goes on opening the valve until the level reaches the set point. Then he stops opening. He recognises that the level goes on rising, but not until it reaches 100mm above the set point does he start closing the valve, which he does in the same methodical, slow and steady way as he opened it. The level is now drops to the set point but does not stop there. It goes on dropping. Ivan repeats his actions several times but never gets to the constant level.

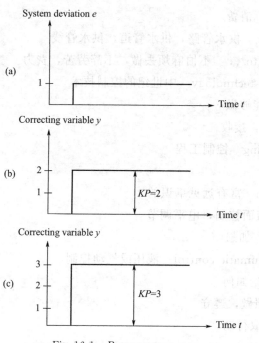

Fig. 16-1　P response curve

5. What was the problem?

Through the slow oscillations the level exceeds or drops below the set point all the time. Ivan behaves like an integral action controller where the correction speed is proportional to the deviation between the set point and the controlled variable. If there is a deviation, Ivan goes on adjusting the valve, never reaching a stable level since the control has an integral-action, see Fig. 16-1.

Part 2 New Words and Phrases

once upon a time　曾经
be in charge of　负责，掌理
water supply　供水系统，水源
a fabulous city　一个很好的城市
advanced age　高龄，老年
social security　社会保险，社会保障
and so on　等等，诸如此类
in addition　另外，此外
make an agreement with　和……达成一致
job description　工作说明
be responsible for　对……负责
water tower　水塔
in consumption　消费
water supply line　供水管路，供水管道，供水管线
take the trouble to...　不怕麻烦去做，不辞劳苦，费力
medieval control technology　中世纪的控制技术
competent to　能够做……
take over　接管，接收
control engineering　控制工程
at length　最后
with great vision　富有远见卓识的
level control　液面控制，电平调节
build up　增进，加强
hydraulic or pneumatic control　液压或气动控制
in any case　无论如何
comply with　照做，遵守
take action　采取行动
as quickly and efficiently as possible　尽量迅速和有效

take place　发生
supply line　补给线，运输线，供电线，军需品供应线路
so far　到目前为止，迄今为止
come up with　提出，想出，赶上
agree upon　对……取得一致意见
start off with　从……开始，用……开始
methodical approach　妥善的处理方法
water level　水位，水平面
set point　设定值
start to do　开始做……
inlet valve　进气阀，进给阀
at the end　最后
go on　继续
not until　直到……才……
all the time　一直
integral action controller　积分作用控制器，积分型控制器，积分控制器
be proportional to　与……成比例
reach a stable level　达到一个稳定的水平

Part 3　Reference Version

第十六课　PID 传说（一）

很久以前，在一个遥远的地方，有一名叫埃米尔的工程师负责为一个美丽城市供水。在他年迈时，他不得不退休。

他将他的退休金和社会保险金等都做了安排。他和法律顾问确定他的遗产留给他三个儿子的其中之一，他们分别是：彼得、伊凡、大卫。但是还没确定是谁。

1. 工作说明

父亲埃米尔一直负责保持当地水塔的水位稳定（实际上为是实现每家每户得到恒压）不受用水量的影响。而这需要根据用水量打开或关上通向水塔的供水线上的三个阀门。

因为用水量差别很大，因此，这不是一项容易的工作。他的儿子们常常来看望埃米尔，不厌其烦地看他的技术活。他们已经学会了这种老式的控制技术，都认为自己有能力胜任这份工作。

在埃米尔要退休时，他的一个亲戚奥斯卡来看望他。奥斯卡学过控制工程。他们详细地讨论控制系统，认为出于长远考虑，应该建立一个手动的、电动的、液压传动的、机械的、液压或气动液面控制系统。

2. 三项要求

该控制系统必须无条件地满足三个要求：

- 在出现干扰（用水量出现差异）时水位会发生变化，负责人应立即迅速采取有效措施来恢复水位。
- 恢复水位所需要的水量需要在任何时候都准确无误。
- 负责人一定要在供水线上保证水位和压力都在一定范围内。

3. 测试负责人

埃米尔即将退休，但是他的儿子们都没有想到办法来满足这三个要求。当地理事局分析了当前的情况，并且商定在艾米尔监督下对他的三个儿子进行技能测试。通过这样一个实证的方式可以找到合适的接班人。

4. 规矩的伊凡

伊凡有条不紊地开始了测验。当他看见水位低于标准水位，下降了100毫米时，他知道用水量增加了。因此，他缓慢地打开进给阀。不久水位下降开始缓慢直到恢复水位。但他仍然不关闭阀门，直到水位达到上限。一切正如伊凡所料，他开始缓慢地关闭阀门。但是最后水位仍旧降到标准水位以下。伊凡多次重复，但仍旧未能保持水位不变。

5. 出现的问题

通过这样有规律地控制阀门，水位上升或下降。伊凡就像一个标准的动作控制器，修正速度与最高水位和受控水量间的偏差一致。伊凡不停地修正偏差，却不能维持稳定的水位。见图 16-1。

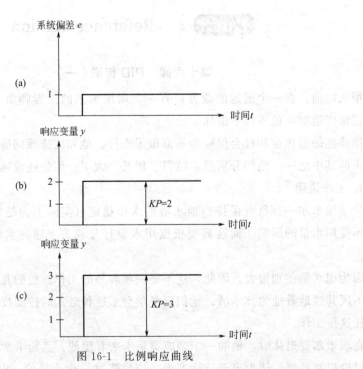

图 16-1　比例响应曲线

Lesson 17 "The Sage of PID"——An Educational Tale for Those Who Would Understand the Concept of PID (2)

Part 1 ▶▶ Text

6. Peter has a sense of proportion

Peter starts his attempt after lunch. As a systematic person he has planned in advance how to do it. It does not take him many seconds to figure out that he has to open the valve 5 turns to balance the level drop of 100mm. On the other hand he is not too keen about getting the level to the set point. As he assumes that the consumption will lower quite soon, the level will rise again. At first there is no signs he is wrong, but after a while he sees that the level is 50mm too high. According to his past experience, this level change fits in with 2.5 turns on the valve to be compensated. He then quickly closes the valve 2.5 turns to keep the pressure constant. At any readable deviation from the set point, he opens or closes the valve as much as he thinks necessary to stabilise the level. Peter's operation is proportional to the deviation between the set point and the controlled variable the same way as aproportional action controller behaves, see Fig. 17-1.

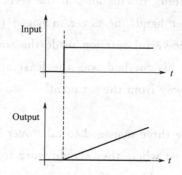

Fig. 17-1 I response curve

7. David-always in a hurry

David had been waiting all day for his attempt. The knowledge of his brothers relative failure had made him plan his job carefully. The result was a complicated method with a lot of work. David studies the speed at which the level drops. If it drops fast, he opens the valve 10 turns as fast as possible and the closes it again just as fast when the level starts to rise again. When he has closed the valve, he notices that the level has deviated from the set point.

David only reacts until a change in the level takes place-the same way as a differential action controller, see Fig. 17-2.

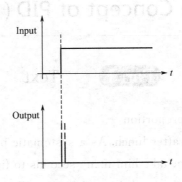

Fig. 17-2　D response curve

8. Emil evaluates his sons

After the tests was finished, Emil discussed his sons' performances with the deputy chief of the councils technical department. They could not decide who the best solution and should succeed Emil. Ivan needed far too long to get back to the set point and his way of doing it had resulted in an unsuitable situation. The continuous change in the valve position resulted in fluctuations in the level and he lacked the ability to dampen them. His only solution was a slow adjustment of valves, but the slower Ivan worked, the longer the citizens would have to wait to get the correct water supply pressure back. Peter succeeded in avoiding fluctuations in the level and the deviations had not been all too big. On the other hand, he never managed to keep the level to the set point. David fast quick and powerful reaction made the water supply act like a whiplash. Despite all his efforts, his method was the least attractive one since the level moved increasingly further away from the set point.

9. PI and PD

As obviously none of the three managed to take over Emil's function all alone, it was decided to make a new test where they should work together.

Technically this was no problem, since there were——as mentioned earlier——three parallel supply lines to the water tower, each with it's own valve. Unfortunately, as the test was to start, David was sick, so Peter and Ivan had to show what they could achieve together. They stuck to their earlier methods. If the level dropped 100mm, Peter opened the valve 5 turns immediately and thereby avoided the level drop. Ivan, at the same time worked slowly and determined to get the level back to the set point, and now he did not have to compensate for the rise in consumption. The only work to do was to finish what Peter had started and correct his carelessness caused by rushing. Ivan actually had even less to do, since there were no fluctuations due to wide-open valves. Together Peter and Ivan acted like a PI controller, their co-operation resulted in an immediate restoring of the water level and a first class result because of Ivan's integral way of behaviour. The consumers were all in all satisfied, but still wanted to see what happened with all three working together before they made any decision. Next time a test was to be made, poor Ivan was ill, so Peter and David had to handle it together. Just as the test started, the water level dropped to the low level mark. Peter tried, sticking to his method, to stabilise the level by opening the valve proportional to the level drop. David immediately opened his valve 10 turns so that the water supply rose to a high flow. He was convinced that Peter was lacking behind and that it was his own quick action which prevented the drop of the level. He was so convinced that he never even checked the deviation compared to the set point. Peter's "measuring" method and immediate reaction stabilised the drop. David, on the other hand, had amplified Peter's influence and avoided a larger deviation from the set point by his "rush" action. In spite of this they never managed to get to the set point even if they operated with a small deviation.

10. PID

At last Emil managed to get all three sons together for the last test. Peter as always opened the valve 5 turns This time too, the level dropped 100mm below set point when the test started. Ivan, according to his habit, started to open the valve to get back to the set point. David immediately opened the valve 10 turns. What happened was this: David avoided the large deviation by his quick acting. Peter handled the interference and Ivan took the deviation away.

Together the three succeeded in finding the ideal solution. The result was that Peter, Ivan and David were employed to take care of father Emil's former job. All the citizens-except one-were happy with the result. The only person not applauding the solution was the council's bookkeeper, who not only had to pay Emil's pension, but also had to pay no less than 3 sets of salaries instead of just one. Everyone else thought it fine that technology should create jobs.

And it is a fine story to tell!

Part 2 ▶▶ New Words and Phrases

systematic person 很有系统性的人
in advance 预先，提前
figure out 解决，算出，想出，理解，断定

on the other hand   另一方面
at first   起先，首先，最初
after a while   过了一会儿
according to   根据，依据
readable deviation   可读性偏差
as much as   差不多，和……一样多
controlled variable   控制变量，受控变量
in a hurry   立即，匆忙
all day   整天
as fast as possible   尽可能快
deviate from   偏离，脱离
deputy chief   副局长，副科长，副处长，副警长，副总警监
technical department   技术部
get back to   回到
continuous change   连续变速
in the valve position   在阀的位置
succeed in   成功，在……方面成功，顺利完成
on the other hand   另一方面
manage to   设法，达成
all one's efforts   全力以赴
further away from   离……更远
all alone   独立地，独自地
decide to   决定
make a test   做实验
parallel supply line   并行供给线
stick to   坚持，粘住
at the same time   同时，然而
determine to   决定
compensate for   赔偿，补偿
due to   由于，应归于
first class   头等，第一流，最高级
make any decision   作出任何决定
drop to   下降到，跌到
rise to   上升到，升迁
be convinced that   确信，信服，相信
compare to   把……比作，比喻为
immediate reaction   即现反应，即时反应

in spite of 尽管，不管，不顾
even if 即使，虽然
at last 最后
take...away 带走，拿走，除去，使消失
take care of 照顾，注意，抚养
be happy with 高兴……事情，对……满意
not only...but also... 不仅……而且……
instead of 代替，而不是……

Part 3 ▶▶ Reference Version

第十七课　PID 传说（二）

6. 有分寸的彼得

有分寸的彼得已经提前计划好了如何开始测验。没多久他就分析出他需要开 5 转阀门来平衡水位，但标准水位不在计划内。如他所料，用水量下降后很快上升。开始没有表现出什么不对，但不久他发现水位高出了 50mm。根据经验，2.5 转阀门后水位应该恢复。他迅速转回 2.5 转阀门以保持水位稳定。在可读性误差范围内他按照预想开关阀门。彼得的操作与最高水位和受控水位间的误差是一致的。见图 17-1。

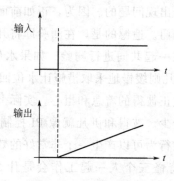

图 17-1　积分响应曲线

7. 着急的大卫

大卫等待了一天，终于轮到他了。两个哥哥的失败让他变得小心翼翼。结果是大量的工作和复杂的方法构成的。他研究了水位下降时的速度。下降速度快时他快速地开 10 转阀门，一旦水位上升，就立即关闭阀门。却发现实际水位已经从设定值偏离。大卫在水位刚一发生变化时就立即采取措施，这和微分控制器的运行方式一样。见图 17-2。

8. 埃米尔的评价

测试结束后埃米尔和当地技术部门的副主管讨论孩子们的表现，仍旧不能够确定

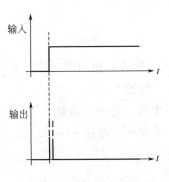

图 17-2 微分响应曲线

哪个解决方案最好及谁能接任埃米尔。伊凡需要太长的时间恢复最高水位，这会产生不好的结果。阀门位置的不断变化会引起水位的不断变化，而伊凡又没有能力控制。伊凡的动作越慢，居民们需要合适的用水压力的时间越长。彼得成功地控制了水位的浮动现象，但是误差太大了。而且，他总是不能让水位和设定值一致。大卫反应快，能够及时处理水位的变化情况，但是因为水位和设定值差别很大，其方法最不可取。

9. PI 和 PD

非常明显，埃米尔的三个儿子都不能够单独胜任他的工作。最终的决定是他们共同配合完成新的测验。

从技术上来说这样是不会出现问题的。因为，正如前面提到过的，三个平行的通向水塔的供水线各自有各自的阀门。遗憾的是，在测验即将开始时，大卫病了，因此彼得和伊凡把两个人的方法结合在一起共同进行测验。如果水位下降 100mm，彼得立即开 5 转阀门来避免水位下降。伊凡则缓慢地采取措施让水位回到设定值，不需要考虑抵消用水量。唯一要做的工作是改正彼得的着急和粗心。实际上，因为阀门的完全打开，水位的波动很小，伊凡的工作最少。彼得和伊凡就像 PI 控制器一样。他们的合作可以让水位恢复得很快，伊凡的有效行为可以产生一个非常好的效果。

居民们都很满意，都很奇怪三个人一起工作会是什么效果。但是在测验要开始时，可怜的伊凡病了。彼得和大卫不得不共同测验。和前面的测验一样，水位开始下降。

彼得用他的方法——根据下降的水位相应地打开阀门——努力地控制水位。大卫立即打开 10 转阀门，水位开始上升。他确定彼得的方法会出现低水位的情况，而他的快速反应可以避免这个现象。大卫确定知道，他不可能检查水位的误差。彼得的测量方法和快速反应稳定了水位。而且，大卫扩大了彼得的方法的效果，避免了更大的误差。尽管如此，他们在小误差范围内没有控制好水位。

10. PID

最后，埃米尔设法让三个孩子共同参加了测验。彼得和以前一样在水位下降 100mm 时开 5 转阀门。伊凡则照旧打开阀门恢复水位。大卫则仍然立即开 10 转阀门。

这次的结果是：大卫的快速反应避免了大的误差；皮特成功处理干扰；伊凡则消减了误差。

彼得、伊凡和大卫共同解决了问题。因此，结果是：彼得、伊凡和大卫共同接管了埃米尔的工作。所有的居民——除了唯一的一个人——都对结果很满意。这唯一的人就是当地的财务人员，他不仅要付埃米尔的退休金，还要付三个人的工资。所有的人都认为拥有技术可以为你提供工作。

这是多么好的一个故事结局！

参 考 文 献

[1] 刘宇. 机电一体化专业英语 [M]. 天津：天津大学出版社，2010.

[2] 姚永玉，常云朋，周丽丹. 机电专业英语 [M]. 北京：人民邮电出版社，2009.

[3] 徐善存. 机电专业英语 [M]. 北京：机械工业出版社，2015.

[4] 张锋，张云龙. 电气自动化技术专业英语 [M]. 北京：北京交通大学出版社，2011.

[5] 童春利. 自动化类专业英语 [M]. 北京：中国电力出版社，2014.